FOXBAT TALES

THE MiG-25 IN COMBAT

MIKE GUARDIA

From *Debrief: A Complete History of US Aerial Engagements, 1981 to the Present* by Craig Brown. Used by permission of Schiffer Publishing. Any third-party use of this material, outside of this publication, is prohibited. Interested parties must apply directly to Schiffer Publishing for permission.

From *Iranian F-14 Tomcat Units in Combat*, by Tom Cooper and Farzad Bishop, Osprey Publishing, used in accordance with UK copyright laws regarding Fair Usage, quoting 800 words or less.

From *Libyan Air Wars – Part 1: 1973-1985*, by Tom Cooper and Albert Grandolini, Helion & Company, used in accordance with UK copyright laws regarding Fair Usage, quoting 800 words or less.

From *Libyan Air Wars – Part 2: 1985-1986*, by Tom Cooper, Albert Grandolini and Arnaud Delande, Helion & Company, used in accordance with UK copyright laws regarding Fair Usage, quoting 800 words or less.

From *Israeli F-15 Eagle Units in Combat*, by Shlomo Aloni, Osprey Publishing, used in accordance with UK copyright laws regarding Fair Usage, quoting 800 words or less.

From *F-15C vs. MiG-23/25: Iraq 1991*, by Tom Cooper and Doug Dildy, Osprey Publishing, used in accordance with UK copyright laws regarding Fair Usage, quoting 800 words or less.

From *MiG-25 'Foxbat;' MiG-31 'Foxhound:' Russia's Defensive Front Line,* by Yefim Gordon, Midland Publishing, used in accordance with UK copyright laws regarding Fair Usage, quoting 800 words or less.

From *Soviet Spyplanes of the Cold War,* by Yefim Gordon and Dmitriy Komissarov, Pen & Sword Books, used in accordance with UK copyright laws regarding Fair Usage, quoting 800 words or less.

Published by Magnum Books
PO Box 1661
Maple Grove, MN 55311

www.mikeguardia.com

ISBN-13: 978-0-9996443-5-5

For Marie and Melanie…

Contents

Contents

Introduction

July 1967: At the Moscow Air Show, the Soviet Defense Ministry unveiled six new state-of-the-art aircraft. From among this lineup of new fighters and interceptors stood the Mikoyan-Gurevich MiG-25—purportedly capable of outrunning and outmaneuvering any aircraft in NATO's inventory.

Yet even before its public appearance in Moscow, the MiG-25 had been a grave concern for Western analysts. Two years earlier, for example, the first operational MiG-25 had flown faster than 1,440 mph on a 1,000-kilometer circuit. During the same flight trial, it climbed to more than 65,000 feet in under three minutes. Later that year, the MiG-25 set a new world altitude record for payload and non-payload flights.

These performance metrics were virtually unheard of in the West.

Indeed, the MiG-25 could fly at speeds in excess of Mach 3 and reach altitudes heretofore deemed unreachable for a tactical fighter. There was little doubt that this new MiG-25, which NATO had code-named the "Foxbat," had broken the airspeed records for East and West. Meanwhile, NATO's intelligence community was baffled by how the Soviets could cobble together such a "masterpiece" of modern engineering.

Existing intelligence on the MiG-25, however, was based on little more than second-hand information. Western analysts knew, for example, that the plane was fast—but the only *public* photos they had seen of the Foxbat were from the Moscow Air Show in July 1967. Four years later, an Israeli F-4 Phantom tried to intercept a Soviet MiG-25 flying reconnaissance for the Egyptians. The engagement was short-lived, however, as the MiG pilot thrust his engines to full afterburner, leaving the F-4 Phantom behind at a speed of Mach 3.2.

Ironically, the US and its NATO allies soon discovered that many of their assumptions about the MiG-25 were wrong. In September 1976, Soviet pilot Lieutenant Viktor Belenko defected from the USSR and flew his MiG-25 to Japan, bringing with him the highly-classified Pilot's Manual. Taking custody of both Belenko and his plane, the US had its first chance to deconstruct the legendary Foxbat. The reality, however, was that this so-called "interceptor" was a poorly-engineered airframe with an oversized motor.

American scientists and engineers quickly discovered that the MiG-25 was never designed for close combat, and the plane was difficult to maneuver. Indeed, its sole purpose was to fly fast. The plane was heavier than expected; its fuel economy was lower than expected; and it was barely maneuverable at lower speeds and at lower altitudes. In fact, the MiG-25 prototypes that had broken airspeed records a decade earlier had been made from different materials—just to break those records. Indeed, these demonstrator prototypes could never have functioned in aerial combat.

Although the MiG-25 (and many of its latter-day stablemates) proved to be a disappointment by NATO standards, it nevertheless enjoyed a long service life in the latter-day Soviet Air Force and several of its allies. The USSR remained the sole operator of the MiG-25 until the 1980s, when it was exported to the Middle East—including Libya, Iraq, and Syria. The MiG-25 saw combat in Syrian service during the Arab-Israeli air campaigns of the 1980s. Almost simultaneously, Iraqi MiG-25s engaged Iranian F-4 Phantoms and F-14 Tomcats during the Iran-Iraq War. Many of these Iraqi Foxbats would later see action against coalition forces during the Gulf War in 1991. Although the Foxbat was usually the loser when pitted against Western aircraft, one Iraqi MiG-25 was confirmed to have downed an American F/A-18 Hornet during the opening stages of Operation Desert Storm.

Today, less than a handful of countries maintain an active fleet of MiG-25s—including Libya, Syria, and Algeria. The Russian Air Force retired the last of its Foxbats in the mid-2010s. Despite its questionable construction, and relatively poor performance in combat, the MiG-25 excelled in its primary role as an interceptor. By design, interceptors are intended to fly fast against unidentified aircraft, and thus have fewer defensive armaments than a typical air-superiority fighter. Whether guarding the Soviet frontier, or flying patrols over the Mediterranean, the MiG-25 Foxbat was a common sight during the latter decades of the Cold War.

From the Arab world to the Iron Curtain, *Foxbat Tales* is the definitive operational and combat history of the MiG-25.

★

Chapter 1

Birth of an Interceptor

The story of the Mikoyan-Gurevich MiG-25 begins in the inaugural decade of the Cold War. As the ideological divide between East and West grew more fervent, so too did their respective methods of intelligence-gathering. The development of long-range reconnaissance aircraft was perhaps the most conspicuous example of the ongoing game

The Mikoyan-Gurevich MiG-25 "Foxbat." A high-speed interceptor that debuted at the Moscow Air Show in 1967, the MiG-25 was purportedly capable of outmaneuvering any jet in NATO's inventory. Although much of the hype surrounding the MiG-25 subsided in later years, its appearance spurred the development of the McDonnell Douglas F-15 Eagle. (Dmitri Mottl)

Profile rendition of the Ye-155, the first operational prototype of what would eventually become the MiG-25 Foxbat (RFI).

of geo-political brinksmanship. Although NATO had kept a close eye on the development of Soviet aircraft throughout the 1950s, none of the Communists' airframes gave cause for concern among the Western intelligence community. Aside from the Tupelov Tu-95 "Bear" and the latter-day MiG-21, NATO had little regard for any aircraft within the Warsaw Pact inventory.

This relative complacency, however, had ended by the dawn of the 1960s. In May 1960, for example, an American U-2 spy plane was shot down by a Soviet surface-to-air missile near Sverdlovsk. This surprise shootdown demonstrated that the US could no longer fly unhindered reconnaissance missions over Soviet territory. Prior to this U-2 loss, NATO had regarded the Soviet Air Defense Forces (PVO - *Protivovozdooshnaya oborona*) as little more than a paper tiger.

To make matters worse, as the US began the first of its abortive experiments with the F-111, the Soviet Union unveiled two new bombers (Myasishchev M-50 "Bounder" and Tupelov Tu-22 "Blinder") and two new interceptors (Ye-152A "Flipper" and Tupelov Tu-128 "Fiddler"). These new planes, although untested in combat, were enough to gain the attention of NATO's frontline intelligence community. To this point, the PVO had been relying on a rapidly-aging fleet of day interceptors—most of which were easily outrun by the current inventory of NATO aircraft. Meanwhile, to strengthen the PVO's intercept capabilities, Soviet design bureaus including Yakovlev, Sukhoi, and Mikoyan-Gurevich submitted prototypes for what they hoped would be the next generation of Soviet interceptors. Yakovlev and Mikoyan-Gurevich respectively developed the Yak-27V and MiG-19SU variants, both of which had rocket-propelled engines.

However, neither aircraft proceeded beyond the prototype phase.

Sukhoi, meanwhile, developed the Su-9 interceptor, achieving operational status in 1959. Its flight range, however, was severely limited, thus rendering the Su-9 impractical as a strategic interceptor.

Because NATO had a growing number of strategic bombers (all of which were capable of delivering nuclear weapons from standoff distances), the Soviet Ministry of

A rear view of the MiG-25 Foxbat as it accelerates to full afterburner. Designed by the powerhouse team, Mikoyan-Gurevich, the MiG-25 was fielded in two variants: an interceptor and a reconnaissance platform. The interceptor was given to the Soviet Air Defense Forces (PVO) as a safeguard against the SR-71 Blackbird and other prowling jets. The reconnaissance variant was delivered to the Soviet Air Force, wherein it saw action in Egypt and during the Soviet-Afghan War. (US Department of Defense)

The twin-mounted engines of the MiG-25. The MiG's engines was arguably its greatest asset. At the time of the plane's unveiling, the MiG-25 was the only tactical interceptor capable of reaching Mach 3. (Vitaly Kuzmin)

Defense ordered the development of a new long-range, high-altitude, high-speed interceptor. By design, these interceptors lacked the customary armaments found aboard modern-day fighters—including autocannons and the normal contingent of close-range missiles. Indeed, their sole mission was to find and destroy NATO bombers and reconnaissance planes, preferably from beyond visual range. These "heavy interceptors," as they came to be known, were never intended for dogfighting.

Development of this next-generation interceptor ultimately befell the Mikoyan-Gurevich design bureau. The true genesis of the MiG-25, however, remains somewhat of a mystery. According to one legend, the story of the MiG-25 began when Chief Designer Artem Mikoyan returned to Russia following his visit to the Paris Air Show in 1959. Having seen the latest and greatest of NATO's aircraft, Mikoyan suggested that his project designer, Yakov Seletskiy, begin researching designs for a new interceptor comparable to the North

American A-5 Vigilante, but powered by twin engines. Official state records, however, indicate that Mikoyan had initiated the design planning before any information was available on the A-5 Vigilante. These records further indicate that sketches were available by 1958, but serious design work did not begin until mid-1959.

These initial designs called for a revolutionary approach. Integrating the desired avionics and weaponry into a suitable airframe would stretch the manufacturing capabilities of the Soviet defense industry. As the new designs were made public, both the PVO and the Soviet Air Force took a keen interest in the project. The former needed a high-altitude interceptor, while the latter needed a new reconnaissance platform.

As it turned out, the Soviet Air Force and PVO had similar requirements for their desired recon and interceptor platforms. For example, both services wanted an aircraft capable of Mach 3 and a service ceiling in excess of 20,000 meters (65,000 feet). Thus, the Soviet government decided to make the forthcoming interceptor a "joint-service aircraft" that could fulfill both roles. Thus, in February 1961 the Communist Party issued a joint directive with the Soviet Council of Ministers, tasking Mikoyan to develop the prototype "Ye-155"—a high-speed, high-altitude aircraft that would eventually become the MiG-25. Two variants of the Ye-155 would be developed concurrently: an interceptor designated Ye-155P (*perekhvatchik*); and a reconnaissance platform designated Ye-155R (*razvedchik*) respectively. On March 10, 1961, Artem Mikoyan formally launched the design/prototype phase of the Ye-155.

Mikoyan's team soon developed three prototype designs— all of which envisioned a high-speed interceptor powered by twin engines. One design featured two adjacent powerplants situated aft of the rear stabilizer, similar to the MiG-19 and the latter-day Ye-152A. Another design featured a stepped-tandem arrangement similar to the I-320 experimental

Above and Opposite. Aerial views of the MiG-25's underbelly. The interceptor variant of the MiG-25 was initially equipped to carry a complement of R-40 air-to-air missiles. The original reconnaissance variant, however, had no defensive armaments. (US Department of Defense)

fighter. The third design utilized two engines paired vertically along the rear fuselage.

Mikoyan selected the first design option because the desired R15B-300 engines had a large diameter, thus rendering a vertical arrangement impractical for the airframe's height and maneuverability. Moreover, the vertical staggering would have complicated the routine maintenance and servicing of the engines. The design team also rejected the idea of placing the engines underneath the wing nacelles due to the potential "thrust asymmetry" that would occur if one engine failed. Simultaneously, Mikoyan–Gurevich determined that the new

plane would not feature the nose-mounted air intakes that had been common among earlier MiGs.

As development of the Ye-155 prototype got underway, Mikhail Gurevich (the second half of the Mikoyan-Gurevich bureau) assumed the project lead as its chief test engineer. Joining him to integrate the on-board avionics and weaponry was Nikolay Z. Matyuk, who had formerly been the project manager for the I-75 and Ye-150 interceptors. Gurevich took lead for the first few years of the project's development, but his advanced age and failing health forced his retirement in 1964, after which Matyuk assumed full responsibility for the project.

During the developmental phase of the Ye-155, designers constructed the airframe largely from stainless steel, accounting for most of the aircraft's tare weight. The engine bays were encased with titanium alloys, while heat-resistant strands of duralumin were used to construct the non-stress bearing airframe components. The steel construction was prescient because it eliminated the fracturing problem that had affected titanium-weld designs found on earlier Soviet and Western aircraft. Moreover, steel was less expensive to acquire and easier to manipulate. Still, the metallurgical overhaul was not without its drawbacks—the steel construction necessitated re-tooling the Mikoyan-Gurevich production facility and, subsequently, the assembly workers had to be re-trained on the attendant technologies.

Originally, Mikoyan intended the Ye-155P to carry two (and later four) K-9M radar-guided missiles. However, as time went on, Soviet munitions designer Matus Bisnovat proposed arming the Ye- 155P with the new K-40 high-altitude, air-to-air missile. The K-40's titanium body made it lightweight and more resistant to kinetic heating at speeds in excess of Mach 4. There were two variants of the missile: the K-40R and K-40T, each operating on different guidance systems. The K-40R was a semi-active radar homing missile, while the K-40T was an infrared heat seeker. Taken together, both missiles would make the Ye-155 a versatile plane since heat-seekers allowed the aircraft to attack from any angle, and the radar-homing system was resistant to enemy countermeasures. At various times throughout the design process, Mikoyan-Gurevich considered arming the prototype Ye-155 with an autocannon, but the idea was eventually scrapped as defensive armaments were considered non-essential for the interceptor/reconnaissance role.

The Ye-155R reconnaissance variant was to be equipped with a state-of-the-art navigation system powered by a digital processor, along with the latest suite of on-board reconnaissance equipment. Aside from the navigational computer, other on-board components included an advanced autopilot system; an air data system; and an air traffic control/identification friend-or-foe (ATC/IFF) transponder.

Because the design team had forsaken the traditional forward air intake concept (and the wing-mounted engine design), Mikhail Gurevich and his team shortened the fuselage

An air-to-air rear view of the Foxbat in flight. For years, the West knew very little about the Soviet's top interceptor. Most of their information came from second-hand sources, or fleeting reconnaissance photos of inconsistent quality. (US Department of Defense)

and reduced the plane's cross-section area, while still leaving enough space to accommodate the projected fuel capacity. Simultaneously, they opted to place the lateral air intakes under the wing, and with horizontal airflow control ramps. The wings themselves would take a trapezoidal form, and the early wind tunnel tests confirmed an acceptable lift-to-drag ratio at speeds in excess of Mach 2. Unlike earlier Soviet jets, this forthcoming MiG-25 would have "shoulder-mounted" wings because "it tied in conveniently with the lateral air intakes and enabled the aircraft to carry large air-to-air missiles, which would have been impossible with a low-wing [or mid-wing] arrangement"—as had been standard on the MiG-15 and MiG-21.

For the interceptor version, Mikoyan-Gurevich selected the Smerch radar system, which had been used aboard the earlier Tu-128. For the Ye-155P, however, the radar received

The MiG-25PD interceptor, dubbed the "Foxbat E" by NATO. Within the Soviet Union, the MiG-25 interceptors were designated by their "P" suffix (e.g. MiG-25P; MiG-25PD; MiG-25PDS, et al). The reconnaissance variants were domestically-identified by their "R" suffix (including the MiG-25R and MiG-25RBS). The MiG-25PD pictured here carries a complement of AA-6 Acrid missiles. (US Department of Defense)

several upgrades, giving it a scan azimuth of 60 degrees and nearly twice the horizontal emission angle of other contemporary Soviet radars. Its vacuum tube innards, however, meant that the radar was prone to break, and expensive to fix.

Still, the interceptor could maintain situational awareness via its integration into the Vozdukh-1 ground-controlled intercept (GCI) system. From ground-based radar stations, these GCI nodes could detect enemy aircraft from several miles away, and guide the defending interceptors onto their targets. GCI was still a developing concept within the Soviet military, but the PVO had been making strides to implement the system for homeland air defense.

The Ye-155R reconnaissance version was radically different from the current fleet of surveillance aircraft in the Soviet Air Force. For instance, the current Yak-25RV could fly high, but the plane was subsonic, and therefore vulnerable to NATO air defenses. Because the Soviet Air Force needed a high-speed reconnaissance platform (and an aircraft presumably capable of penetrating enemy air defenses unhindered), Soviet air commanders expressed great interest in the new recon prototype.

Defensive countermeasures aboard the reconnaissance version included chaff dispensers, infrared heat decoys, radar warning receiver, and an electronic countermeasure suite to jam enemy transmissions. The avionics suite aboard the recon variant included a ground-mapping radar, a Doppler speed and drift sensor system, along with an RSBN-2S Svod short-range radio navigation system. The navigation equipment could function either with an analogue processor or a Plamya-VT flight computer, one of the first digital airborne computers in Soviet history.

Multiple versions of the reconnaissance suite were proposed. All versions could be fitted with a photo adapter, a cockpit voice recorder, and an optical tracker. The reconnaissance version could also be equipped with the interchangeable SRS-4A and SRS-4B Signals Intelligence (SIGINT) packs. By and large, the mission equipment proposed for the Ye-155R was similar to that carried by other Soviet reconnaissance aircraft of the day.

The proposed reconnaissance suites, however, quickly came under criticism from the Soviet Air Force, who labeled the system "inefficient." Thus, in March 1961, the Air Force revised its operational requirements and submitted a list of potential targets that the Ye-155R prototype would be expected to identify and engage. These targets included: "missile launch pads, ammunition depots, naval bases and harbors, ships, railway stations, airfields, command, control, communications and intelligence (CI) centers, soft-skinned and armored vehicles, and bridges." Moreover, the reconnaissance suite had to identify these targets within a few

A side view of the MiG-25P interceptor. Following its debut and registered airspeed records, Western analysts were baffled by how the Soviets had cobbled together such an impressive aircraft. Prior to the appearance of the MiG-25, none of the aircraft from the Warsaw Pact had caused much concern among NATO's intelligence community. (US Department of Defense)

hundred meters of ground truth, in order to facilitate proper engagement from follow-on strike aircraft. The intelligence gathered from the recon variant would then be transferred via data link after flying over the targets.

The Soviet Air Force further asked Mikoyan to consider upgrading the reconnaissance version with the Puma radar, the FARM-2 adapter for capturing photos from the radar screen, a camera programming module, and an RV-25 radio altimeter. Essentially, the Soviet Air Force wanted three sub-variants of the reconnaissance Ye-155R—a signals intelligence (SIGINT) variant; a photo intelligence (PHOTINT) variant; and an electronics intelligence (ELINT) variant.

Mikoyan-Gurevich granted their request, but with several modifications. For example, the design team had determined that a reconnaissance jet stood virtually no chance of penetrating NATO's air defenses at medium altitudes. With minimal cloud cover between 32,000 - 65,000 feet, the Ye-155R would be a proverbial "sitting duck." Thus, the reconnaissance variant would have to perform its missions in supersonic flight at higher altitudes. This necessitated an upgrade to the on-board cameras' focal lengths.

While the mission equipment for the recon variant was being finalized, Mikoyan-Gurevich compared the current inventory of Soviet reconnaissance equipment against the American items salvaged from the wreckage of the U-2 spy plane that had been shot down in May 1960. After some analysis and reverse-engineering, the design team made further recommendations to approve the Ye-155's long-range surveillance capabilities. Meanwhile, the design work for the Ye-155P interceptor and the Ye-155R Reconnaissance variant continued and, by the fall of 1963, the first operational prototype was nearly complete.

Re-designated the "Ye-155-R1," the reconnaissance prototype took its first flight on March 6, 1964. The reconnaissance prototype had some unique characteristics, including a 600-liter wingtip fuel tank on either side of the airframe. However, the flight test determined that these tanks

A MiG-25RB takes off from Voronezh Malshevo Air Base in Russia, 2012. (Alex Beltyukov)

were unstable and the sloshing fuel was causing vibrational stress to the wings. Hence, the fuel tanks were promptly removed. The recon variant also had attachments for movable canards which could regulate the aircraft's pitch at higher speeds. The Ye-155-R1 was quickly followed by five successively-numbered reconnaissance prototypes (R2-R6), each of which featured minor upgrades and modifications to the airframe.

Meanwhile, the interceptor prototype, whose working designation was now "Ye-155-P1," made its first flight on September 9, 1964, piloted by Pyotr Ostapenko. Aside from its forward fuselage, the P1 different little from its proto-reconnaissance stablemates. Indeed, the camera assembly in the nose of the Ye-155R was replaced with a fire-control radar housing. For the initial test flight, two outboard pylons carried dummy K-40 missiles. A second prototype, the Ye-155-P2, arrived at the flight test facility a year later. The P1 was among several MiG-25 prototypes to complete the manufacturer's flight testing and claim new airspeed records. However, the P1 did not participate in the official State acceptance trials.

Because the State trials were evaluating different parameters for the two Ye-155 variants, the Soviet government decided that both versions would be parallel-tested by separate teams. Several of the Ye-155 sub-variants participated in the stakes. In doing so, these prototype MiG-25s claimed a number of impressive airspeed records. The test pilots pushed the aircraft beyond its expected limits, and these record-setting flights took the West by surprise.

To achieve these new airspeed records, Mikoyan-Gurevich took three existing prototypes—Ye-155-P1; Ye-155-R1; and Ye-155-R3—and reconfigured each airframe to make it lighter, nimbler, and to decrease drag. Under the classification system set by the Federation Aeronautique Internationale (FAI), the international governing body for air sports, the Ye-

A MiG-25RB of the Russian Air Force touches down after completing a training sortie, February 2012. Following the breakup of the Soviet Union, the Russian Air Force was the MiG-25's largest legacy operator. Today, the MiG-25 remains in service with only a handful of nations, but most of the existing airframes are in deteriorating condition. (Alex Beltyukov)

155 was a class C1 (III) aircraft—i.e. a jet-powered land plane with no maximum take-off weight. The first record claimed by a Ye-155/MiG-25 prototype was an airspeed record with take-off weights of 1,000 and 2,000 kilograms. Under these parameters, on March 16, 1965, test pilot Aleksandr Fedotov reached speeds in excess of 1,440 mph along a 1,000-kilometer circuit. Two years later, on October 5, 1967, Fedotov reached an altitude of 98,350 feet with a 2,200-lb payload. That same day, test pilot Mikhail M. Komarov averaged 1,852 mph over a 500-kilometer closed circuit with no payload.

Aeronautical feats of that magnitude were unheard of in the West.

Having satisfied the State evaluators over the course of its exhaustive flight testing, the Soviet government approved the Ye-155 prototypes for series production in 1969. With the stroke of a pen, the Ye-155 officially became the MiG-25.

A Soviet airman stands beside a MiG-25, highlighting the enormity of plane itself. The Foxbat was larger and heavier than most interceptors of its day. (US Department of Defense)

Low-rate initial production of the MiG-25 began in 1969 at the Gorky aircraft factory. More than 1,186 units were built by the time production ended in 1984. The first production-variant interceptor was dubbed the "MiG-25P." At the time of its arrival, the American SR-71 Blackbird reconnaissance aircraft was completing its final flight tests. Soviet intelligence had determined that the SR-71 was faster than most frontline jets in either NATO or the Warsaw Pact. Because the MiG-25P was the only Soviet aircraft capable of countering the SR-71, the PVO High Command requested the new MiG-25Ps to guard the major industrial and military centers along the Soviet borderlands.

In April 1969 the PVO's 148th Combat & Conversion Training Center at Savasleyka received delivery of the first MiG-25Ps. These initial MiGs were delivered primarily for training and evaluative purposes. Pilots and ground crews needed familiarization with the new airframe and its associated maintenance. As Savasleyka was close to the aircraft plant in Gorky, factory technicians were always nearby to help troubleshoot any problematic airframes. Both the pilot manuals and the servicing manuals were being refined during the Savasleyka testing, but the pilots and technical crews soon mastered the MiG-25P without issue.

The initial MiG-25P evaluations also demonstrated that the plane could fire its R-40 missiles in both "head-on" and "pursuit" modes at targets up to 65,000 feet and speeds in excess of Mach 2. These metrics proved that the MiG-25P could, theoretically, intercept and destroy the SR-71. Shortly after the MiG-25P's unveiling, NATO designated it the "Foxbat-A." The "Foxbat" reporting name would stay with the MiG-25 (and its variants) for the remainder of its service life.

Full-scale production of the MiG-25P began in 1971 at the Gorky plant. The series-production MiG-25 differed little from the early-production series that had test-flown at Savasleyka. Indeed, aside from its re-contoured fins and improved landing gear, the series-production model was essentially the same aircraft. Part and parcel to the MiG-25P was the Smerch-A1 radar, around which the entire weapons control system had been built. According to Soviet aviation historian, Yefim Gordon: "The radar could scan the airspace

A two-seat MiG-25RU trainer takes to the sky. Both the interceptor and reconnaissance MiG-25s were available in two-seat trainer variants—respectively designated the MiG-25PU and MiG-25RU. (Dmitriy Pichugin)

A MiG-25 trainer at Zhukovsky International Airport, 1995. (Rob Schleiffert)

and track aerial targets either autonomously or using ground inputs relayed via the Vozdukh-1 GCI system."

Over the next five years, the MiG-25P received several upgrades to its avionics suite. The biggest overhaul, however, came on the tail of Lieutenant Viktor Belenko's infamous defection. After flying his MiG-25P to Japan, defecting to the United States, and surrendering his MiG-25 to Western analysts, the Soviet Defense Ministry found itself in an undeniable predicament. The poor quality and craftsmanship of the base-model MiG-25P interceptor had been exposed.

The aura and mystique of the Foxbat was gone.

Now that the hype of the MiG-25 had faded, it was clear to the PVO that the MiG-25 needed an upgrade to its avionics and weaponry, else the plane would be rendered useless in its combat role. The Soviets thus devised a new Weapons Control System for integration into all forthcoming MiG-25s, and for retrofitting into the existing fleet.

The Smerch-A radar was thus replaced by the Sapfeer-25, a quasi-continuous emission system derived from an earlier model found aboard the MiG-23. A new Infrared Search &

Track (IRST) system would be integrated with the radar, thus making the entire system less susceptible to enemy jamming. Alongside the new airborne radar, the PVO likewise developed an improved GCI system, including an aircraft-mounted receiver that was supposedly impervious to enemy countermeasures.

Complementing the new avionics were an assortment of upgraded missiles, each having an extended "kill range" made possible by the improved seeker heads and higher-capacity storage batteries. These new missiles included the R-60 and APU-60-2, both of which could be affixed to the outboard pylons.

On November 4, 1976, the Council of Ministers and Communist Party Central Committee issued a communique tasking Mikoyan to build the new-and-improved MiG-25. It would soon be designated the "MiG-25PD." The "D" suffix stood for "dorabotannyy"—meaning "upgraded." This new-and-improved model was to enter production immediately, as well as a simultaneous conversion program to upgrade the existing MiG-25Ps to the "PD" configuration.

Outwardly, the MiG-25PD differed only slightly from its predecessor. The front fuselage was modified slightly to accommodate the new Sapfeer radar. The skin panels had likewise been re-contoured and the maintenance access ports had been relocated according to the new internal equipment layout. The IRST fairing was affixed to the underside of the nose and, much like the reconnaissance variant of the MiG-25, the "PD" interceptor could accommodate a 5,300-liter drop tank.

The Soviet Air Force and PVO remained the sole operators of the MiG-25PD until the 1980s, when an export version was developed for the Middle East and North Africa. The export variant was somewhat of a hybrid between the MiG-25PD and the earlier MiG-25P. For example, the MiG-25PD

An artist's rendition of a US pilot intercepting a Soviet MiG-25 at high altitude. Even after the West discovered that the MiG-25 was not a highly-maneuverable nor particularly well-engineered jet, NATO air forces nonetheless considered it a threat due to its speed and radar capabilities. (US Department of Defense)

export combined the long-nosed body of the domestic PD-model with the Smerch-A2 radar system of the P-variant. However, like the domestic MiG-25PD, the export version could also carry the R-60 missile. As with most military exports, some of the MiG-25's advanced avionics were downgraded—just in case the recipients later became enemies of the Soviet Union. The export MiG-25PD was delivered to

Libya, Iraq, Syria and Algeria. It was these foreign-supplied MiG-25PDs that saw the most action throughout the Foxbat's history.

The reconnaissance variant of the MiG-25—dubbed the "MiG-25R"—rolled off the assembly line in parallel to its P-model stablemate. As the MiG-25P interceptor completed its first operational trials, so too did the MiG-25R. To accommodate its role as a surveillance and reconnaissance aircraft, the MiG-25R featured four A-70M cameras for regular PHOTINT missions, and an NE-10 topographic camera with a 1300mm lens. Collectively, this camera suite could capture images from altitudes as high as 72,000 feet. Unlike the interceptor variant, the MiG-25R carried integral fuel tanks in its fins to extend its operational range. Its defensive equipment, however, consisted of little more than an electronic countermeasures pack—which could jam enemy radar and transmissions.

Per the Soviet Air Force's mandate, the first few production aircraft arrived at the Lipetsk aerodrome for conversion training, a reconnaissance assessment, and operational evaluation trials. Most of the new MiG-25Rs, however, went to the Moscow Military District - specifically to the Guards Independent Reconnaissance Air Regiment at Shatalovo Air Base. It was this regiment that began the formal service testing program for the MiG-25R in the summer of 1969.

As the MiG-25Rs were delivered to the Soviet Air Force, numerous test pilots—including veteran MiG tester Aleksandr Fedotov—rendered assistance to the frontline air regiments for their conversion training. Several units, however, lacked computer technicians who were qualified to work on the MiG-25R. Rapid changes were thus made to the curriculum at the Soviet Air Force Engineering Academy, with some students being reclassified as MiG-25 technicians during mid-semester. Still, the initial round of flight and serviceability tests were completed and the MiG-25R officially became the principal reconnaissance aircraft of the Soviet Air Force.

Later, all MiG-25Rs were upgraded to the MiG-25RB standard. The "RB" suffix denoted its adaptation into a multi-role strike aircraft. However, the true genesis of the RB reconnaissance-bomber varies depending on the source. Most sources point to the RB's birth coming in the wake of the Six-Day War in 1967, wherein Egypt and Syria (staunch allies of the Soviet Union) were defeated by the Israeli Defense Force. Following their victory in the Six-Day War, Israel began conducting strike missions over Egyptian territory in an effort to maintain air superiority. However, after an Israeli airstrike destroyed a transformer station in Cairo, knocking out power to the entire city, Egyptian President Gamal Nasser approached the USSR, "requesting technical and military assistance for the air defense, reconnaissance, and strike missions" against Israeli targets. Because Egypt and Syria were important geo-political allies, Soviet Defense Minister Dmitriy Ustinov, decided to send the MiG-25R to the Middle East as both a reconnaissance plane *and* a tactical bomber. The reasoning behind this newfound dual-role was that if the MiG-25R could carry flares for night reconnaissance missions, it could likewise accommodate air-to-ground munitions.

Test pilot Stepan Mikoyan, however, tells a different story of how the RB-variant came to be. According to Mikoyan, the Minister of Aircraft Industry, Pyotr Dementyev, was so inspired by the MiG-25R's navigation suite, that he hypothecated using the MiG-25R as a high-altitude bomber. But whether it was Nasser's request, or Dementyev's inspiration, Mikoyan-Gurevich began converting its "pure reconnaissance" MiG-25R to a dual-role strike aircraft in 1969.

Since the MiG-25 had no bomb sight, the bomb would be released automatically by the on-board navigation computer

Two examples of Ukrainian Air Force MiG-25s. Following the breakup of the USSR, many of the remaining MiG-25s were passed along to the successor states. Ukraine was one of the foremost non-Russian operators of the MiG-25 until the Ukrainian Air Force retired the jet in the late 2000s. (Alamy)

The derelict remains of a MiG-25 at the Khodynka Airfield in Moscow. Since their withdrawal from frontline service in Russia, several of these Foxbats have been sent to the scrapyards. (Aeroprints)

as the aircraft approached its target. To accommodate this computerized bomb deployment, the design team installed a bomb trajectory calculation software into the navigation computer.

In March 1970, after the Soviet Air Force was confident that the plane could function as a bomber, test pilot Aviard Fastovets made the first bomb drop from a MiG-25R at an altitude of 65,620 feet, travelling at 1,552 mph. Deploying a bomb from that airspeed and altitude was, by itself, another world record. Soviet Air Force pilots Aleksandr Bezhevets and Nikolay Stogov later took over the flight testing for the MiG-25R bombing evaluations and conversions. The testing process, however, was not without its setbacks. Defects in the onboard electrical system became an early nuisance, along with synchronization lapses in the navigational system. During an April 1970 test flight, with Aleksandr Bezhevets at the stick, the navigation computer malfunctioned and, as the MiG reached supersonic speed, the actuating cartridges overheated, causing a premature bomb release. To prevent any further un-commanded releases, the bomb racks were relocated to a cooler area underneath the fuselage; thereby reducing the payload capacity to 4,410 lbs.

Concurrent with these bombing tests on the MiG-25R, the Gorky aircraft plant began producing the first official MiG-25RB. Aside from its reconnaissance mission, and based upon feedback from the initial bombing tests, the MiG-25 would have equipment to facilitate all-weather day and night bombing attacks. For its anticipated bombing missions, the RB-variant could accurately drop bombs from altitudes of nearly 69,000 feet at airspeeds in excess of Mach 2. At full armament, however, the RB's combat radius fell to 403 miles as opposed to the 571-mile radius of the pure reconnaissance variant. After releasing its bombs, the MiG-25RB could maintain an airspeed of at least Mach 2.25 while performing its getaway maneuvers. After extensive testing, the MiG-25RB became operational in December 1970. It was the first of the Foxbat's recon-strike variants. Further iterations of the MiG-25RB included the MiG-25RBK and MiG-25RBS, both of which featured minor upgrades to the airframe and avionics.

An export version of the MiG-25RBK was developed almost in tandem with the export MiG-25PD. Like the interceptor, the exported MiG-25RBK was a slightly-downgraded version of its domestic counterpart—"featuring less-sophisticated guidance and weapons control systems." More than thirty airframes were delivered to Algeria; five to Libya; eight to Iraq; eight to Syria; and six to India, who had recently become a Soviet client state. The Soviet Defense Ministry had also delivered three MiG-25s to the Bulgarian Air Force, who later exchanged them for MiG-23 fighter-bombers. All told, the Bulgarian Air Force had no practical use for a high-speed reconnaissance aircraft.

Throughout its service life, there were two training variants of the MiG-25: the "MiG-25PU" and "MiG-25RU," respective trainers for the interceptor and reconnaissance versions. The PU-variant had no armaments or radar, as it was simply designed to familiarize the pilot with the basics of handling the Foxbat. The RU, however, featured the basic navigational/recon system found upon the standard R-variants.

The MiG-25P and MiG-25R (and their numerous variants) saw extensive service both within and beyond the Soviet Union. The USSR was the largest historical operator of the aircraft—maintaining several hundred MiG-25s in their inventory until the end of the Cold War. After the dissolution of the USSR, these MiG-25s were passed along to the various successor states—the Russian Air Force being the primary operator. Other latter-day Soviet recipients included Armenia, Azerbaijan, Belarus, Georgia, Kazakhstan, Turkmenistan, and Ukraine. Most of these former Soviet republics have since retired their MiG-25s. The Russian Air Force had intended to keep the Foxbat in limited service as a reconnaissance aircraft until 2020, having announced that they would upgrade the existing fleet of MiG-25s to bring them in line with the current stable of fourth-generation aircraft. In December 2013, however, the Russian Air Force retired the last of its MiG-25Rs, citing that the airframe was too expensive to upgrade or maintain.

Beyond the former Eastern Bloc, the MiG-25 saw its most extensive use in North Africa and the Middle East. Of these latter-day foreign operators, the Iraqi Air Force was the most prolific, although most of their Foxbats were destroyed during the Iran-Iraq War and the subsequent Gulf War of 1991. The Indian Air Force made substantial use of the MiG-25, but kept the plane a tightly-held secret until its retirement in 2006. Today, in the Middle East and North Africa, only Algeria, Libya, and Syria maintain an active fleet of MiG-25s, although in limited numbers and in various states of operability.

Chapter 2
The Great Escape

B y all outward appearances, Lieutenant Viktor Ivanovich Belenko was a model Soviet citizen. A cradle Communist and devoted MiG pilot, Belenko seemed destined for a long and prosperous career in service to the USSR. Underneath, however, Belenko had grown disenchanted with Communism and he was chafing under the heavy-handed repression of the Soviet regime. By the age of 29, he had reached his breaking point. He wanted out of the USSR.

Born on February 15, 1947 in Nalchik, Russia, Belenko grew up in a typical working-class family, raised primarily by his father and stepmother. At the age of 20, he enrolled in the Armavir Higher Military Aviation School of Pilots, where he graduated in 1971. Commissioned as a pilot in the Soviet Air Defense Forces (PVO), he initially served as an instructor at the Stavropol flight school before being trained as a MiG-25 pilot. In 1975, at his own request, he was transferred to the Far East sector of the PVO's operational area—standing at the edge of both Alaskan and Japanese airspace. Orders assigned him to the 513th Fighter Aviation Regiment of the 11th Separate Air Defense Army at Chuguyevka Air Base.

By this point, however, Belenko had grown tired of the Soviet system. The living conditions at Chuguyevka Air Base were dismal; the flight crews suffered from low morale; there was no oversight or accountability from the chain of command; workdays were unregulated; and families were neglected. When Belenko raised these issues to the regiment's political officer, however, and offered suggestions for improvement, he was derided and hastily silenced. To make matters worse, his wife Lyudmila had grown dissatisfied with her life as a military spouse. She informed him of her intent to divorce him and to move with their young son, Dmitry, to her parents' home in Magadan, Russia.

On the morning of September 6, 1976, Viktor Belenko prepared for what should have been another routine flight over the Sea of Japan. Unbeknownst to his superiors, however, Viktor had smuggled the MiG-25's highly-classified technical manual into the cockpit (a strictly-forbidden practice).

Belenko was defecting from the Soviet Union.

When he ignited the engines of his MiG-25P at 6:40 AM, no one suspected that his defection was imminent. Taking flight from the airfield, Belenko deliberately fell behind the lead elements in his formation. When he failed to return to Chuguyevka at the appointed hour, air traffic control attempted to contact him. Receiving no response, Belenko's superiors assumed he had crashed, and quickly dispatched a group of fighters to locate his crash site. Little did they know, however, that Viktor Belenko was already in Japanese airspace.

After falling behind his flight leader, Belenko descended his MiG to an altitude of 100 feet to avoid detection from Soviet radar. Skimming just above the surface of the sea, the wayward pilot vectored his MiG-25 on a course for Hokkaido, Japan. In a curious twist, however, it was later confirmed that US intelligence agents were on the ground in Hokkaido, and were expecting his arrival. It remains unknown whether Belenko contacted these agencies on his own, or if they

Official ID photograph of Lieutenant Viktor Belenko, a MiG-25 pilot serving in the Soviet Air Defense Forces (PVO). Disenchanted with communism and the heavy-handed Soviet regime, Belenko shocked the world by flying his MiG-25 from eastern Russia to Japan, whereupon he announced his defection and surrendered his Foxbat to Japanese and Western authorities. (Central Intelligence Agency)

bribed him via the spy network operating within the Soviet Union. Whatever the case, at 1:10 PM, Japanese radar detected the signature from Belenko's MiG and scrambled two F-4EJ Phantom fighters from Chitose Air Base near Sapporo. These Japanese F-4s, however, were new aircraft, having entered service with the Japan Air Self-Defense Force (JASDF) in 1974. As such, their pilots hadn't truly mastered the plane's intercept capabilities, and the F-4's radar wasn't really optimized for that role anyway. Taken together with the low visibility that morning, neither F-4 could locate Belenko's plane as it approached the Japanese coast.

Belenko, meanwhile, was having his own navigational difficulties. Indeed, the MiG-25's navigation system could not adequately guide the aircraft during extended low-level flight. Thus, Belenko had to rely on nothing more than "dead reckoning" and his own intuition. His map of Hokkaido had only shown Chitose Air Base, and he anticipated landing there after being intercepted by the JASDF. However, by his own estimate, he was rapidly approaching the Japanese coast and he had not seen a single JASDF aircraft.

At approximately 1:30 PM, Belenko breached Japan's territorial airspace. By now, he realized that his MiG had vectored off course from the JASDF air base and began looking for an alternate landing site. Running low on fuel and needing to land quickly, he located Hakodate Airport in southern Hokkaido.

The landing itself, however, was no simple matter.

Belenko circled the airport three times before landing and, upon his descent, nearly hit a departing Boeing 727. To make matters worse, the runway at Hakodate was only 6,000 feet— too short to accommodate a MiG-25. Despite deploying his plane's drag parachute, the front tires of Belenko's landing gear burst upon touchdown, and the aircraft skidded some 790 feet off the end of the runway. Kicking up asphalt and

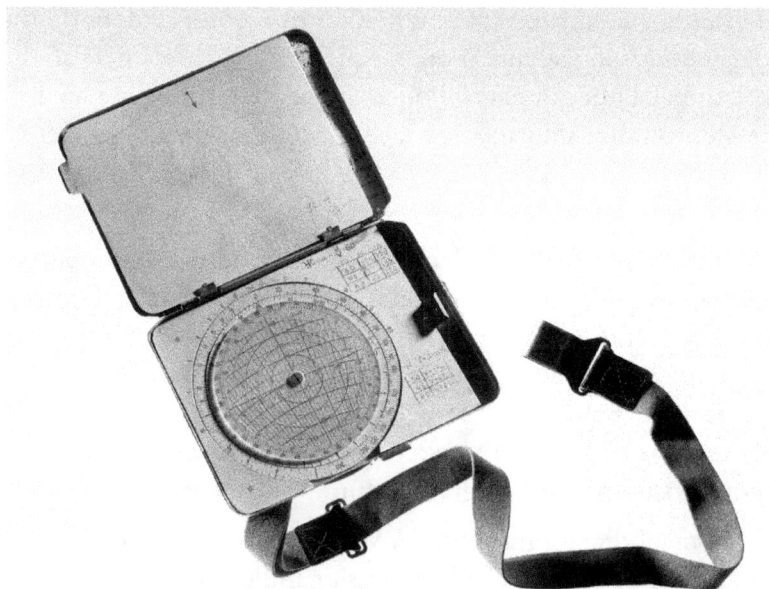

Viktor Belenko's leg-strap notebook with flight data. It was one of only three items he brought with him on his flight—the other being his military identification and the MiG-25's manual. (Central Intelligence Agency)

plowing through the dirt, the MiG-25 ground to a halt near the airport's localizer antenna. By the time Belenko shut down his engines, he had only 30 seconds of fuel remaining.

Shaken from the rough landing, but otherwise unharmed, Belenko regained his bearings in time to see a horde of Japanese civilians descending upon his plane. Naturally, the local citizens were curious to see the Soviet fighter that had just invaded their hometown airport. As the crowd began to swarm the MiG and take pictures of the aircraft, Belenko drew his service pistol and fired two warning shots into the air.

The Japanese civilians dispersed, but the local police did not.

Having witnessed the final approach and the ungraceful

landing, Hakodate Air Traffic Control contacted both the Japanese Self-Defense Force and the local precinct of the National Police Agency. Police arrived on the scene at 2:10 PM, promptly shutting down the airport, and taking Belenko into custody.

Within a few hours, Belenko's landing had become a media sensation. The police arrested him for violating Japanese airspace and for illegally discharging a firearm. While in custody, however, he made a statement to the press and requested political asylum in the US. On September 7, Belenko was moved to Tokyo, and the following day, the US announced that it had granted him political asylum.

Soviet officials, meanwhile, were astounded by the whole affair. Nothing in Belenko's dossier indicated that he would become such a high-profile defector. He had been a loyal member of the Communist Party, a talented pilot, and he had not expressed any pro-Western sentiments. The Soviet embassy requested an interview with Belenko and demanded his extradition to Moscow. On September 9, 1976, a Soviet envoy in Tokyo met with Belenko and unsuccessfully tried to convince him to return to the USSR. Following that brief unsuccessful meeting, Belenko boarded a Northwest Orient Airlines flight to the United States.

Meanwhile, the Japanese Ministry of Justice transferred control of the impounded MiG to the Ministry of Defense. The Japanese Self Defense Force, however, was now on high alert—and were none too happy about the presence of a MiG-25 in their own country. Fears abounded that there would be a Soviet attack or an attempt to recover the aircraft by Soviet agents. After Belenko's landing, the 11th Division of the Japanese Ground Self Defense Force (JGSDF) deployed 200 troops to the Hakodate Airport, armed with Type 61 tanks and 35mm anti-aircraft guns. The Japanese Maritime

Self-Defense Force (JMSDF), meanwhile, deployed vessels running cyclic patrols in the Sea of Japan and along the Tsugaru Strait between Hokkaido and Honshu. At the same time, local JASDF aircraft, including the F-4EJ Phantoms from Chitose Air Base, ran 24-hour patrols over Hokkaido to intercept any incoming aircraft.

After its arrival in Hakodate, the MiG-25 had been cordoned off and partially covered. Belenko's crash had rendered the plane unairworthy, and the Hakodate Airport had no facilities to accommodate the storage and deconstruction of the Soviet plane. Thus, on September 25, it was partly disassembled and crated aboard a US Air Force C-5A Galaxy, where it was flown from Hakodate Airport to Hyakuri Air Base north of Tokyo. As the C-5A departed Hokkaido, a banner on the plane read, in Japanese: "Goodbye people of Hakodate, Sorry for the trouble."

Arriving in Honshu, the MiG-25 was quickly descended upon by a team of Japanese and American technicians, all of whom were eager to see the inner workings of Russia's top interceptor. Belenko had left the technical manual with the aircraft, expecting that he would demonstrate it for the US Air Force. Given the current condition of the plane, however, and Belenko's sequestration to the US, the team of technicians simply began the process of deconstructing the MiG.

This was the first time that Western experts were able to inspect the aircraft up close. What they discovered, however, was that the MiG-25 was not the formidable fighter they had anticipated. The JASDF, for example, had heretofore classified it as a fighter-bomber. They were surprised (and relieved), however, to learn that the Foxbat was merely an interceptor. Throughout the deconstruction process, the American-Japanese team found that the MiG-25's construction was "crude," but conceded that the steel airframe could withstand high temperatures and could be repaired with minimal maintenance skills. They were less impressed by the onboard

Japanese authorities descend upon Belenko's MiG-25, shortly after its touchdown at the Hakodate Airport in Hokkaido, September 1976. (Getty Images)

Smerch radar, whose components were outdated. Western analysts did, however, praise the Smerch for its double wavebands—"which made it virtually jam-proof—something no US fire control radar featured at the time."

Although the deconstruction of Belenko's MiG had revealed an aeronautical "paper tiger," the plane nonetheless had characteristics that the West could not ignore—namely its speed and radar capabilities. To this end, the US continued developing and fielding the F-15 Eagle, an air superiority fighter that had initially been designed as a counterweight to the MiG-25.

While the team of analysts continued their work at Hyakuri Air Base, the Soviet government pressured Japan to return the MiG. At the time, a CIA report determined that "both countries seem anxious to put the problem behind them" and that the USSR could not afford "setbacks in Soviet–Japanese economic cooperation." Thus, on October 2, 1976, after completing the American-led analysis and deconstruction of the MiG-25, the Japanese government laid out its plan to return the airframe in a series of crates shipped from the port of Hitachi.

The Japanese had deliberately crated the MiG within forty boxes to obscure the fact that their analysts had tampered with the plane. When the Soviet recovery team arrived at Hitachi, they demanded the crates be opened for inspection.

Indeed, they wanted to ensure that nothing was missing from their wayward MiG.

The Japanese government, however, gave the Soviet team only a few hours, hoping that it wouldn't be enough time for a thorough inspection. On November 15, the disassembled MiG was finally shipped from Hitachi to Vladivostok. Upon its return to the USSR, however, Soviet experts were quick to discover how much the West had gleaned from the MiG-25. For example, they determined that the Americans had run a static test of the engines and had studied the aircraft's heat signature. The Soviets could also tell that the Americans had tampered with the avionics, including the Smerch radar. "Not knowing how to operate the equipment," said Gordon Yefim, "the Americans had damaged some of it and had to make hasty repairs...foreign fuses and resistors were discovered in the radar."

Concluding their inspection, the Soviets determined that nearly 20 pieces of the aircraft had gone missing—including the onboard film footage of the flight to Hakodate. Moscow billed the Japanese government $20 million for the cost of the MiG. The Japanese, in turn, responded with a $40,000 bill for their crating services and the cost to repair the runway at Hakodate. Neither bill was ever paid.

Then-CIA Director George HW Bush called the MiG-25 an "intelligence bonanza." Not only had Belenko delivered a functional Foxbat, he also provided valuable information on the state of the Soviet Air Force, PVO, and the forthcoming MiG-31 "Foxhound," which would eventually replace the MiG-25.

Upon his arrival in the US, Belenko was debriefed for five months by the CIA and the Department of Defense. Receiving a stipend from the CIA, Belenko became a highly-regarded consultant for the aerospace industry and US military. Meanwhile, he mastered English, became a US citizen, and gradually adapted to life in America. He later wrote of his experience in the Soviet military, and his defection, in the book *MiG Pilot: The Final Escape of Lieutenant Belenko*, co-authored with John Barron in 1980.

While residing in the US, Belenko married a music teacher from North Dakota, with whom he fathered two sons. However, Belenko has never officially divorced his Russian wife, Lyudmila. Although he rarely appears in interviews, Belenko states that he has been content with his life in the United States. He has publicly stated that Americans "have tolerance regarding other people's opinion. In certain cultures, if you do not accept the mainstream, you would be booted out or might disappear. Here we have people, you know, who hug trees, and people who want to cut them down—and they live side by side!"

As Belenko started his new life in America, the Soviet government cobbled together a PR campaign for "damage control" in the wake of his defection. At first, the Soviets claimed that Belenko had simply gotten lost on a routine aerial mission and had to make an "emergency landing" at Hakodate. The Soviet media also claimed that he had been drugged by the Japanese and imprisoned in America against his will. The state-run media later amended its statement, claiming that Belenko's landing occurred "under unclear circumstances," and called the Western media's characterization of the event a "propaganda campaign." The Soviet Ministry of Foreign Affairs officially dismissed the notion of a defection—calling it "a lie, from beginning to end." Soviet diplomats insisted that Belenko had made his statements under duress, and claimed that the Japanese had refused to let them speak to Belenko privately while he was in Tokyo.

On September 28, the Soviets held a press conference in Moscow featuring Belenko's wife and his stepmother, both of whom tearfully asked for his return—calling him a "patriot of the Soviet homeland" and a "loving husband and son." Concurrently, a spokesman for the Ministry of Foreign Affairs stated that America's role in the MiG-25 incident had been "tantamount to the forcible separation of the Belenko family." Statements from the Soviet government and the Belenko family were broadcast far and wide by the USSR's press outlets, in hopes that the news would eventually reach the defector. For the next several months, the official version of the incident in the Soviet press did not change: "Senior Lieutenant Belenko made an emergency landing in Hokkaido due to lack of fuel, was criminally abducted by the Japanese authorities under the dictation of Washington." For years afterward, however, the Soviet Union repeatedly printed false stories about Belenko. Many of these stories were contradictory—one claiming he had been killed in a car accident; one claiming he had been executed; while another claimed he had simply been "arrested."

Family ties and media campaigns notwithstanding, the bigger problem for the Soviet Union was that the MiG-25 had been compromised. To make matters worse, Belenko's statements had given NATO a primer on the forthcoming MiG-31. With a technical readout of the MiG-25, and basic information on the MiG-31, the West could now develop appropriate countermeasures. Motivated by the loss of Belenko and his aircraft, the Ministry of Defense began working at breakneck speed to develop the new MiG-25PD,

Upon being apprehended by Japanese police, Belenko requested political asylum in the US. During his debriefing, he provided information on the then-forthcoming MiG-31 (pictured here), touted to be an eventual replacement for the MiG-25. (Alex Beltyukov)

which entered production in 1978. The PD would become the operational basis for the MiG-31. The following year, the PVO began upgrading its operational MiG-25Ps to the PDS standard. By 1983, the PVO's combat capabilities had vastly improved thanks to the arrival of the MiG-25PD, MiG-25PDS, and the new MiG-31 interceptor, which by then had achieved operational status.

In the wake of Belenko's defection, the Japanese Self Defense Force began re-evaluating its air defense posture. Indeed, Belenko's MiG had violated Japanese airspace—yet the ground-based radars in Hokkaido had not properly tracked him, and the JASDF fighters failed to intercept him. The JASDF subsequently purchased a fleet of Grumman E-2 Hawkeye Airborne Warning and Control System (AWACS) aircraft, and later developed their own licensed version of the F-15 Eagle, with better look-down radar capabilities than the aging F-4 Phantom.

Meanwhile, on the other side of the Iron Curtain, a government committee visited the Chuguyevka Air Base to investigate Belenko's publicized claims of the poor living conditions. The investigators were astonished by what they found. Even by Soviet standards, the living conditions at Chuguyevka (and the unit's command climate) were abysmal. The Soviet government began an immediate overhaul of the air base—improving the facilities and quality-of-life programs.

Even after Belenko's defection, however, the MiG-25 remained a critical piece of the Soviet air defense network (until the arrival of the MiG-31), as the Foxbat was still the only aircraft capable of intercepting the SR-71 Blackbird. For its part, the SR-71 was a perennial nuisance over the Baltic Sea. For instance, during the popular uprisings in Poland in the 1980s, NATO feared a possible Soviet invasion. Thus, to assess the disposition of Soviet forces along the western border, the Americans sent the SR-71 aloft over the Baltic, whereupon the MiG-25PD and PDS frequently intercepted the prowling Blackbirds.

★

Chapter 3
MiGs of the Mediterranean

While the Free World clamored for more information about the elusive Foxbat, the MiG-25R made its combat debut in the skies over the Middle East. Having close ties to the Arab World, the Soviet Union took stock of Egypt's defeat during the Six Day War of 1967. Indeed, the Soviets were shocked (and begrudgingly impressed) by the speed and ferocity with which the Israelis had vanquished their Arab neighbors.

Given the ties of friendship between Egypt and the USSR, it came as little surprise when, in January 1970, President Nasser visited Moscow, soliciting help to rebuild and retrain the Egyptian armed forces. Israeli air-ground synchronization had overwhelmed the Egyptian and Syrian forces, and Nasser wanted something to improve his air defense capabilities.

Without much ado, the Soviet government granted Nasser's request.

Indeed, as early as February 1970, Egyptian military personnel began training in the USSR and took delivery of the latest in Soviet equipment. Meanwhile, Soviet military advisors poured into Cairo by the hundreds. By April, several PVO surface-to-air missile (SAM) batteries and tactical fighter units had arrived in Egypt to protect targets of strategic importance—including the Aswan dam and the seaport of Alexandria.

The Soviet advisors also took part in planning operations aimed at freeing the Sinai Peninsula, which Israel had annexed following the Six Day War. Under these anticipated battle plans, the Egyptian Army was to cross the Suez Canal and proceed onward into the Sinai. In any case, careful observation was vital, since the Israelis had built up a strong protective line along the Suez—the so-called Bar-Lev line.

By 1971, the Soviets realized that another Arab-Israeli conflict was on the horizon. By the same token, however, the Soviets knew that even with their best advisors and equipment, the Egyptians were in no shape for another war against Israel. Direct involvement from the Soviet military was out of the question, as it would certainly trigger World War III. Thus, the Soviet government chose to send a special reconnaissance team of MiG-25Rs to Egypt.

At the time, the MiG-25 program was in a state of limbo. Sadly, the initial fielding and conversion training had not gone as smoothly as planned. To make matters worse, the PVO's commander-in-chief of aviation, General Anatoly Kadomtsev, had been killed in a fatal crash while piloting a MiG-25P in April 1969. All told, the Soviet Air Force and PVO were growing wary of the MiG-25, and both services doubted whether the Foxbat could be the viable recon-interceptor it was touted to be. Amidst this growing doubt, however, the Vice-Minister of Aircraft Industry, Aleksey Minayev, proposed sending the MiG-25 to the Middle East, wherein the plane could prove itself under real-world conditions.

The Soviet military approved Minayev's plan, but realized that it would have to be conducted cautiously. The Soviet Union could only spare a handful of aircraft—indeed, larger quantities could attract attention from the West, and likely be

A reconnaissance-variant MiG-25, similar to those flown by the Soviet pilots of Det 63 during their deployment to Egypt in 1971. Amidst the growing tension between Israel and Egypt following the Six Day War, the Soviet Air Force deployed a detachment of MiG-25s to conduct reconnaissance flyovers of the Sinai Peninsula. (Alex Beltyukov)

construed as "direct intervention." Also, there would be no advantage in sending the MiG-25P to the Middle East. The P-model interceptor could only be effective when flying at longer ranges and at higher altitudes, and the geographical confines of the Middle East simply weren't conducive to these aerial maneuvers. Moreover, considering the close proximity of the combatants (Egypt, Israel, Syria, and Jordan), they were more likely to use lightweight air-superiority fighters than heavy interceptors. Thus, the Soviet Air Force elected to send four reconnaissance MiG-25R aircraft to the Middle East. These recon MiGs (piloted by Soviet aviators and remaining under Soviet control) would fly at higher altitudes and provide timely intelligence of Israeli dispositions.

The Soviet task force included seven pilots, all of whom were closely guarded by Egyptian commandos from the moment they arrived in country. These seven pilots were, arguably, the best tactical aviators in the Soviet Union. Their ranks included Vladimir Gordiyenko, who had flown nearly every production model of the MiG-25 and had been a flight instructor on the aircraft. Commanding this coterie of seasoned aviators was veteran test pilot Colonel Aleksandr Bezhevets—"a man renowned for his resolve and command skills…second to none in knowing the MiG-25, having flown the first Ye-155 prototypes back in 1965."

For this mission, the Soviet Air Force selected two "pure reconnaissance" MiG-25Rs and two MiG-25RB reconnaissance/strike aircraft. For their PHOTINT duties, each aircraft could accommodate two different camera sets, along with interchangeable SIGINT packs. At first, none of the Soviet pilots knew they were going to Egypt; their destination was kept a secret until the last possible moment. During their flight physicals, however, the pilots had to endure a series of examinations to assess their fitness for operating within "hot and dry climatic zones"—suggesting either North Africa or the Middle East. After the group confirmed that their mission was to "extend international help" to the Republic of Egypt, they stood ready to deploy in the fall of 1970.

However, following President Nasser's death in September 1970, the Soviet government postponed the deployment. At first, it seemed that the new Egyptian leader, Anwar Sadat, was more interested in *negotiating* with Israel than resuming an armed conflict. However, when Sadat confirmed that he was willing to take back the Sinai by force, the MiG-25 deployment went ahead as planned.

In March 1971, the first of the Soviet contingent arrived in Cairo. In an effort to save time and money, the Soviet Air Force dismantled the four MiG-25s and airlifted them into Egypt via An-22 transports.

But the process of getting these disassembled MiGs aboard

the airlift was no simple task.

For even with its wings, engines, and tail fins removed, the MiG-25 still couldn't fit into the An-22's cargo bay. Indeed, by a margin of only a few inches, the mainwheels of the landing gear got stuck on the cargo door. To create more space, one of the handlers suggested temporarily attaching MiG-21 wheels to the main gear struts. These MiG-21 wheels would be strong enough to carry the weight of a stripped-down MiG-25, while giving the plane the smaller profile it needed to fit within the cargo hold.

The MiG-25 task force was officially designated the "63d Independent Air Detachment" (Det 63) and stationed at Cairo-West Air Base. To maintain the secrecy of the mission, all Det 63 members wore Egyptian uniforms. The detachment reported directly to Colonel-General Vasilii Okunev, the top Soviet military advisor in Egypt. Ahead of the Soviets' arrival, the Egyptians had built four hardened aircraft shelters (HAS) to accommodate the incoming MiGs and their technical crews. Under the cover of every HAS, the Soviet technicians carefully reassembled the Foxbats within a matter of days.

Their work, however, was confounded by sporadic bombardment from the Israeli Air Force. Indeed, Israeli aircraft attacked Cairo-West several times throughout Det 63's tenure, causing the air defenses at the base to be reinforced by multiple SAM batteries. The HAS bays containing the MiG-25s were further guarded by a platoon of ZSU-23-4 self-propelled anti-aircraft guns, all of which were manned by Soviet crews. Soviet soldiers also provided ground security for the base, building machine gun emplacements and barbed wire fences. After fully re-assembling the MiG-25s, the Foxbats were then relocated into revetments previously occupied by Egyptian Air Force Tu-16s.

Although "security" and "secrecy" were the governing principles of the Soviet task force, the local Egyptians had no such discretion. As one aviation historian described it: "Egyptian officers never gave security a second thought, and having them participate in mission planning and support meant that the Israelis were aware of the group's plans almost before the meeting adjourned." For all their good intentions, good faith, and gracious hosting, the Egyptian officers simply couldn't keep their mouths shut. These loose-lipped officers, however, paled in comparison to the Egyptian media. Mere days after touching down in Cairo, Det 63 was exposed by the local newspaper, *Al-Akhram*, carrying the headline: "New aircraft at Cairo-West Air Base." The newspaper referred to the MiG-25 as the "X-500," but the accompanying pictures left no doubt that these planes were indeed Soviet MiGs. Frustrated by their host nation's cavalier attitude on security, Det 63 was simultaneously impressed by how quickly the Israelis had obtained information on their planned sorties. To prevent any further security breaches, however, Colonel-General Okunev decided that *all* Det 63 operations (including planning and maintenance) would be conducted exclusively by Soviet personnel.

But the bigger problem, it seemed, was mitigating the long arm of Israel's air defenses. Soviet planners developed special air routes to avoid Israeli aircraft, along with evasive maneuvers to escape the ever-present SAM batteries. Det 63 launched its first flights over Egyptian territory in April 1971. During this time, the MiG pilots tested their cameras and navigational computers while finalizing their mission profiles.

Pilot Vladimir Gordiyenko flew the first mission of Det 63. During his sortie, however, the SIGINT pack recorded that his MiG-25 had been detected by *three* different radars—an Israeli station in the Sinai; a US Navy destroyer in the Mediterranean; and a British surveillance radar in Cyprus. Aside from learning the location of enemy radar sites, however, these training sorties allowed the pilots to see the

The camera suite of the MiG-25RB. With its powerful lenses and climate-controlled apparatus, these camera suites produced high-quality PHOTINT wherever they were used. (RFI)

more inviting aspects of the Egyptian countryside. In order to properly calibrate the MiG-25's onboard instrumentation, the pilots had to follow their predetermined routes very closely. Normally, pilots would follow their prescribed routes by using landmarks as checkpoints. Since there were virtually no landmarks in the featureless desert, however, the pilots used the pyramids in the Valley of the Pharaohs as landmarks, prompting them to call their missions "guided tours."

After nearly a month of conducting flybys and training sorties, Det 63 was ready for its first combat mission. The launches themselves were a painstaking process, with extra steps added for security. For instance, the local Egyptians (and even air traffic control) would not be advised of a pending MiG takeoff, lest some carefree Egyptian announce that a sortie was taking flight. The pilot would thus start his engines while the aircraft was still in the hangar, then run the necessary systems checks, and secretly taxi to the runaway. The MiGs would then unexpectedly take-off, usually causing panic amongst the air traffic controllers at Cairo-West. Of course, the Egyptians were none too pleased by the recurring "surprise" take-offs, but the Soviets accepted their hosts' ire as the cost of maintaining operational security.

The recon Foxbats operated in pairs, with both pilots maintaining strict radio silence and carefully following the designated flight plan as entered into the automatic flight control system. Flying in pairs increased the pilots' probability for mission success. Should one aircraft go down, the other pilot could report the crash site, thus facilitating a search-and-rescue effort. Mission profiles dictated flying over enemy territory at maximum speed for nearly 40 minutes, often at altitudes in excess of 60,000 feet. However, the combined airspeed, altitude, and time in flight, took a heavy toll on the MiGs. The engine's temperature peaked at more than 608°F, while the aircraft's skin registered at 577°F—strains that required cyclic and heavy-duty maintenance. Thanks to the MiG-25R's recent engine upgrades, though, "virtually all sorties could be flown at maximum thrust."

Meanwhile, the Israelis were vexed by the MiG-25. Time and again, the Foxbats penetrated Israeli airspace, outpacing every fighter that tried to intercept them. Because the Soviet pilots maintained radio silence during their sorties, the Israeli Air Force had little choice but to linger outside Cairo-West Air Base, hoping to intercept the MiG-25s as they departed the runway.

But even these ambush tactics failed.

For as soon as the Israeli fighters closed in, they were immediately intercepted by Egyptian MiG-21s flying top cover for the Foxbats. While these MiG-21s flew interference, the Soviet MiG-25s would accelerate beyond Mach 2, leaving the Israeli fighters behind to fend off the swarming Arab MiGs.

Flying at airspeeds of Mach 2.5, no other aircraft in the Middle East could keep pace with the MiG–25. It was just as well, because the MiG–25 reconnaissance variants were unarmed. Aside from the Foxbat's incredible speed, the pilots marveled at the airplane's heat tolerance. In fact, one pilot recalled that, at higher Mach speeds, the glass canopy of the cockpit would get so hot, it would burn his fingers to the touch, even though his hands were fully-gloved. As the MiG–25 approached the designated target area, its camera suite would automatically begin photographing strips of land on either side of the aircraft. To prevent damage from the intense heat of supersonic flight, the camera bay sat within an air-conditioned chamber with an ambient temperature variance of no more than 12.6°F. To facilitate clear photography at such high airspeeds, the onboard camera shutters opened and closed at a comparable rate. To compensate for any unanticipated camera movements, the Foxbat carried special adapters with movable prisms, allowing for continuous focus and clear resolution regardless of the airplane's pitch, roll, or yaw. Aside from taking pictures of Israeli installations and troop movements, these recon MiG–25s pinpointed Israeli radars sites along with Command & Control nodes.

Upon their return to Egyptian airspace, and on final approach to Cairo-West, the Foxbats were met by more MiG–21s for escort to the air base. These MiG–21s would maintain their holding pattern over the airfield until the Foxbats were safely within their hangars and their engines had been shut down. The Israeli Air Force, meanwhile, fed up with losing so many of their aircraft to the SAM ring surrounding Cairo-West, gave up trying to attack the base.

Still, the Arab-Israeli conflict continued.

Although the Israelis had diverted their attention away from the air base, they had now begun targeting the SAM sites around the city. Two of these SAM batteries were destroyed, along with the Soviet crewmen manning them.

This prompted the Soviets to take additional security measures for Det 63. In October 1971, for example, the Soviets built special underground hangars for the MiG-25s. Fitted with all the necessary maintenance and communication assets, these underground bunkers could sustain a direct hit from a 1,000-lb bomb with no damage to the structure itself or the airplane inside. Under the new security measures, the MiG-25s' maintenance and pre-flight checks would be conducted underground, with the aircraft leaving the bunker only before take-off.

The MiG-25s flew only two sorties per month. Their flight plans took them over the entire Suez Canal region and across the Sinai Peninsula. Upon their return to Cairo-West, the MiGs yielded several-hundred yards of film, providing valuable intelligence to the Egyptian war effort. These high-quality photographs, often snapped from altitudes over 65,000 feet, clearly showed the locations of Israeli command centers and troop formations. The quality of the film was so good, in fact, that it even showed areas covered by camouflaged netting. The MiG-25s' SIGINT package, meanwhile, revealed the location of a critical jamming facility, along with air defense radars and various SAM sites.

Throughout the fall of 1971, Det 63 continued its successful flyovers. They continued to push the mileage of their recon flights and, by that winter, the Foxbats were flying within range of the Gaza Strip. Still, no jet in the Israeli Air Force could climb as high or fly as fast. The MiG-25 repeatedly outran the F-4 Phantoms and Mirage IIIs that were sent to hunt it. At best, the Israeli pilots would only catch a fleeting glimpse of the Foxbat before it disappeared beyond the horizon. The Israeli SAM batteries, meanwhile, fared no better than their winged counterparts. Armed with US-built Hawk missiles, the SAM batteries could only reach altitudes of about 40,000 feet. The MiG-25's radar warning receiver could detect when the plane was being tracked by an enemy radar, but the pilot would simply turn on his jammer and carry on with his mission. These "deep penetration flights" continued into the spring of 1972. Israel lodged several complaints to the United Nations, but to no avail.

The Israeli Air Force, meanwhile, tried to come to grips with the elusive Foxbat. At one point, a spokesman for the Israeli Defense Force stated that they had clocked the MiG-25 travelling at more than *Mach 3.2*. Although none of the Soviet pilots significantly deviated from their mission profiles, they did occasionally push the speed envelope. For instance, while attempting to escape from a flight of F-4 Phantoms, Aleksandr Bezhevets accelerated his MiG to more than triple the rate of its prescribed Mach limit. During another mission, Soviet pilot Captain Krasnogorskiy reached a top speed of Mach 2.75. Despite these dangerous speeds, however, the ground crews and technicians were relieved to discover that the airframe had sustained no damage.

Overall, the new MiG-25s had proven themselves reliable in high-tempo reconnaissance missions. Still, as with any aircraft, the desert Foxbats had their share of problems and operational hiccups. On one occasion, pilot Nikolay Stogov suffered a double-engine flameout, turning his MiG-25 into a heavy glider that would likely crash-land. As the dying Foxbat began to decelerate, Stogov radioed for help. Air traffic control gave him two options: return to base immediately or land at the nearest Egyptian air base. Seconds later, however, Stogov's engines miraculously recovered and Stogov proceeded on with his mission. Upon landing, the MiG's ground crew determined that the "stall" had been caused by a momentary disruption to the fuel system. Luckily, the plane's electronic "engine control system" had somehow corrected the problem and reignited the engines.

After one of Aleksandr Bezhevets' missions, his main landing gear strut failed to lock down during his final approach into Cairo-West. Thinking quickly, Bezhevets tried to manipulate his aircraft into a two-point landing. Touching down at 180 mph, he tried to keep the aircraft's weight off the main strut for as long as he could. Inevitably, however, the main strut collapsed and the MiG careened onto its wingtip, scraping along the runway until it skidded to a halt. As it turned out, Bezhevets had executed the landing so gracefully, that the MiG suffered only minor damage to the wingtip, which was quickly mended. Some sources claim that his MiG-25 was repaired in-house at Cairo-West, and was soon back in action. Other sources claim that, Bezhevets' MiG was returned to the Soviet Union for repairs and a substitute MiG-25 was sent in its place.

After several months at Cairo-West, the original cadre of Det 63 redeployed to the USSR in April 1972. The MiGs, however, stayed in Egypt, awaiting a new contingent of Soviet Air Force pilots to continue the reconnaissance missions over Israel. But even with their new pilots, the budding MiG-25s still had their share of mechanical mishaps. During one mission, the cockpit glazing for the Foxbat flown by AY Yashin failed during his deceleration at high altitude. The cockpit began to depressurize, which would have rendered Yashin unconscious (and left for dead), but his pressure suit remained operational and his auxiliary oxygen system kicked in, enabling Yashin to land his MiG without incident.

In total, the Soviet MiGs flew nearly two dozen sorties over Israeli territory. The photographs taken by the recon Foxbats provided critical intelligence, all with high-quality resolution. For the Egyptian military, the high resolution of the aerial photography was a godsend—their own MiG-21RFs had outdated cameras with a narrow field of vision, thus diluting much of the critical detail needed to assess the photographs.

Throughout Det 63's service on the Egyptian front, only one sortie was flown by a single aircraft. The Soviets preferred

An underbelly view of a Libyan MiG-25PDS. This Libyan Foxbat is armed with a complement of R-40RD and R-60MK missiles. The R-40 was a medium-range, air-to-air missile, known in the West as the "AA-6 Acrid." The R-60 was a shorter-range, infrared missile also known as the "AA-6 Aphid." (US Navy)

to operate in pairs, but the operational tempo this day was such that it necessitated a solo flight. Flown by Aleksandr Bezhevets, he maneuvered his lone MiG-25 over the Mediterranean, skirting the boundary of Israel's territorial waters. While maintaining his flight pattern off the coast of Israel, Bezhevets activated his camera suite while banking the aircraft into a sharp turn—an innovative reconnaissance technique to widen the camera's field of view. His maneuvering, combined with the clear air and good lighting, enabled a higher shutter speed which yielded excellent photographic results. But according to his mission profile, Bezhevets wasn't supposed to fly that close to the Israeli border. "However, the navigation specialists had forgotten about the high salinity of the Mediterranean and failed to make corrections to the Doppler speed and drift sensor inputs when programming the navigation computer. As a result, the navigation error amounted to several kilometers...and the aircraft flew directly over the border for three nautical miles." This border-tracing flyby alarmed Israeli military leaders as it demonstrated the weaknesses in their air defense network.

The MiG-25's success in the Middle East bore testament to the plane's resiliency and reliability. Indeed, these reconnaissance missions had erased many lingering doubts of the MiG-25's reliability. And by the end of 1972, the Foxbat had been fully-integrated into the Soviet Air Force and PVO.

In time, however, diplomatic relations between Egypt and the Soviet Union began to sour. The Egyptian Air Force offered to buy the MiG-25 for their own squadrons, but Moscow promptly denied their request. Tensions rose as the Soviet military delegation continued to exclude Egyptian personnel from their planning and operations. Moreover (and perhaps out of spite), Egyptian troops began conducting exercises uncomfortably close to the MiG-25 hangars. Finally, in July 1972, President Sadat expelled all Soviet military advisors, thus ending Det 63's operations in the Middle East. After some closeout negotiations with Egyptian leaders, the Soviet Air Force airlifted the MiGs in the same manner they had arrived—disassembled and stowed aboard An-22 cargo jets. Although the Israelis had seen fleeting glimpses of the Soviet MiGs, they had never managed to shoot one down. Hence, they couldn't prove Soviet involvement and Moscow maintained plausible deniability.

Soviet-Egyptian relations remained cool until October 1973, when Arab forces initiated the Yom Kippur War. At first, Egyptian ground and air forces held the initiative, penetrating the Bar-Lev line and driving deep into Israeli territory. But just as they had done in the two previous Arab-Israeli conflicts, the IDF quickly turned the tables. Within days, the Israeli counteroffensive had pushed the Arab forces back across the Suez Canal. With his back to the wall, and running out of options, Sadat reluctantly turned to the Soviet Union for help.

Thus, on October 19-20, the first new MiG-25RBs arrived in Egypt, landing at Cairo-West. The new task force was labelled "Det 154," which included pilots, engineers, technicians, and factory representatives. This veritable MiG-25 posse was commanded by Lieutenant Colonel Vladmir Uvarov, who had flown with Det 63 the previous year. But as Uvarov and his comrades soon discovered—"the situation was very different from the previous deployment, with Israeli tanks advancing on Cairo at an average of 10 kilometers (6.2 miles) per day. Cannon fire could be heard in Heliopolis, a suburb of Cairo, in the morning hours." As the MiGs prepared for duty at Cairo-West Air Base, Uvarov devised a contingency plan to fly his planes back to the USSR in case the Israelis occupied the city. If the aircraft could not escape in time, Uvarov and his men would simply torch the Foxbats to prevent them from falling into Israeli hands, and Soviet personnel would be evacuated to Libya.

The hype of the new mission, however, was short-lived. A few days after the MiGs arrived in Cairo, Sadat entered negotiations for a cease-fire. By the time hostilities ended on October 25, 1973, only one MiG-25 was flying. But considering that the Israelis had now captured Soviet-built equipment from the Egyptians (including the S-75 air defense missile), Det 154 decided to send a pair of MiG-25s on a recon sortie before the cease-fire took effect. According to the mission profile, one Foxbat was to reconnoiter the Suez Canal, while the other flew over the Sinai proper. The reconnaissance mission was a success, and both pilots returned to base without incident. Their pictures, however, revealed the deplorable state of Egyptian ground forces. In some places along the battlefront, entire Egyptian brigades were being bested by Israeli *platoons*. Following the truce, the MiGs of Det 154 stayed in Egypt for another year, returning home in late 1974.

Although Yom Kippur marked the last Soviet involvement in an Arab conflict, the combined missions of Det 63 and Det 154, validated the MiG-25 as reconnaissance platform, and renewed the Soviets' confidence in the plane. As a result, newer versions of the R-series variants were developed until the MiG-25's production ended in 1984. Taking note of the MiG-25R's success in Egypt, the PVO grew more accepting of their own P-series interceptors, and kept it in frontline service until well after the fall of the Soviet Union.

Syria

The Syrian Air Force was among the first to receive the export variants of the MiG-25PD and MiG-25RB, along with two double-seat trainers. At its peak, the Syrian Air Force operated sixteen PD interceptors and eight RB reconnaissance frames. These Syrian MiGs were assigned to squadrons at the Shayrat, Tiyas, and Dumayr air bases. During their frequent clashes with the Israeli Air Force, Syrian MiG-25s often took flight to intercept Israeli F-15s and F-4 Phantoms. During these engagements, however, the MiG-25 was the recurring loser.

Since the late 1970s, Syria had been a perennial participant in the Lebanese Civil War, a regional conflict involving Maronite Christians, the Palestinian Liberation Organization (PLO), and eventually Israel. Throughout Israel's involvement, the F-15 squadrons flew countless air patrols along the Lebanese border. It was during these regular flyovers when, in the summer of 1979, Israeli F-15s downed several Syrian MiG-21s in the skies over Lebanon. As tension escalated, the Syrian Air Force began flying closer to the Israeli border, hoping to bait the IDF into a protracted aerial battle. As these close encounters continued along the Lebanese border, Israeli F-15s scored more aerial victories against the antiquated

MiG-21s—downing four in September 1979, and confirming an additional three kills in 1980.

By this time, however, the Syrian Air Force had grown increasingly frustrated with the MiG-21. Sworn to vengeance, they acquired newer variants of the MiG-23 and MiG-25 for the next aerial showdown. The Israelis, meanwhile, were also rebuilding. By 1981, they had received more F-15s along with their first delivery of the new F-16 Fighting Falcon.

Although emboldened by their string of victories, the Israeli Air Force still had no indication of how their F-15 would fare in combat against the MiG-25. Although the Eagle had been built in response to the MiG-25 and its stable mates, NATO's intelligence community knew little about the Foxbat beyond its observable performance metrics. Even after Lieutenant Belenko's defection, Western analysts could not conclusively determine how their own fourth-generation fighters would fare against the Foxbat and its latter-day contemporaries. However, on February 13, 1981, any doubts about the Eagle's viability against the MiG-25 were quickly put to rest.

That afternoon, an Israeli F-4 Phantom went aloft, conducting reconnaissance of the Lebanese border when a Syrian MiG-25 vectored to intercept. To this point, Israeli F-4s had been running regular reconnaissance missions over the borderland, where they were frequently targeted by marauding MiG-21s and MiG-23s. These recon variants of the F-4 Phantom never went aloft without a fully-armed, multi-layered Combat Air Patrol (CAP) nearby. But even without a fighter escort, the typical Israeli F-4 could climb faster than the MiG-21 and easily outmaneuver a MiG-23.

Today, however, this F-4 would be targeted by a MiG-25.

Leading the F-15 CAP in support of the F-4's recon mission was Lieutenant Colonel Benny Zinker—

commanding officer of the Israeli F-15 squadron. Because weather conditions were so poor that day, neither Zinker nor his wingmen had anticipated the enemy to interdict the reconnaissance flight. However, the F-15's radar soon populated with the signature of an enemy plane.

General David Ivry, the Israeli Air Force Commander, recalled that "the Syrian fighter was flying an interception course, accelerating very fast." From the relative speed of the radar blip, he could tell that the incoming fighter was either a MiG-23 or a MiG-25. Zinker and his wingmen were flying over the Sea of Galilee when they were alerted to the Syrian fighter. Simultaneously, IDF ground control ordered the F-4 to abort its mission while Zinker vectored his F-15 to meet the lingering bogey.

Acquiring the bandit on his radar, Zinker fired an AIM-7 Sparrow missile from a distance of 25 miles, and launched a second Sparrow only moments later. After a few more moments, Zinker launched a third AIM-7 missile and watched in delight as the amorphous bandit exploded in a flash of light. Zinker saw debris from the enemy fuselage falling to the Earth, but he could not identify whether the bandit had been a MiG-23 or MiG-25.

By now, however, Zinker had wandered into Syrian airspace—and made a hasty retreat before any other MiGs could arrive on the scene. Two weeks later, the Syrian government confirmed that the lost aircraft had indeed been a MiG-25. It was the first time in history that a Foxbat had been bested by enemy fire.

Five months later, on July 29, 1981, Israeli F-15s once again proved their mettle against the MiG-25. After a fragile cease-fire had been declared between Syria and Israel, the latter stated they would still be flying reconnaissance missions over Lebanon's border. Israeli Prime Minister, Menachem Begin,

Another Libyan Foxbat. This photo was taken from a US Navy F-14 Tomcat flying from the USS *Coral Sea* in 1985. (US Navy)

defended the recon flights, stating: "We have to carry out the overflights to know what is going on and to find out where there are the PLO's bases…" Essentially, Israel claimed that it was a matter of preventative defense—lest the PLO use the ceasefire as a pretext to re-arm their border camps.

On July 29, another Israeli F-4 took flight on a reconnaissance mission when two Syrian MiG-21s and two MiG-25s scrambled to intercept. Almost on cue, a flight of F-15s vectored to ward off the Syrian bandits. Within moments of arriving on the scene, the F-15 piloted by Major Shaul Simon acquired radar lock on the leading MiG-25, firing an AIM-7 missile. The Arab Foxbat exploded into a massive fireball—a sight that sent his now-rattled wingmen into retreat. Two of the F-15s pursued the MiG-21s that were fleeing eastbound towards the Syrian border. Unable to close with the Syrian bandits, however, the F-15s soon gave up the chase and vectored back towards Israel.

Just then, the surviving MiG-25 (who had been waiting patiently for the F-15s to break off their pursuit), fired two missiles at the departing F-15s from beyond visual range. The Israeli pilots easily evaded the incoming missiles and turned to re-engage the devious MiG. The Foxbat pilot, however, burned a hasty retreat across the sky, leaving the Eagles behind as he accelerated to supersonic flight. When Syria threatened to shoot down any further reconnaissance flights, Prime Minister Begin replied: "It is easier said than done, because today we have shot down a MiG-25."

Following another cease-fire in July 1981, neither the Israeli nor Syrian Air Force saw much action until May 1982. In their latest round of hostility, PLO forces initiated a 12-day artillery attack on Galilee in northern Israel. By the end of the bombardment, sixty Israeli civilians had been killed. But as the Israeli government contemplated its response, their ambassador to the UK, Shlomo Argov, was brutally shot in the head during an assassination attempt on June 3, 1982. Argov survived the attack, but the gunshot left him paralyzed for the remainder of his life. Israel concluded that the assassination attempt had been part of a PLO conspiracy in combination with the attack on Galilee. The following day, Israel declared the start of Operation "Peace for Galilee"—a full-scale invasion of Lebanon.

On June 6, 1982, seven Israeli divisions—including 60,000 troops and 500 tanks—crossed into southern Lebanon. The IDF had planned the invasion as a three-pronged attack—with air and ground forces deploying along the coastline, the central mountains, and an area known as the Bekaa Valley near the Syrian border. Occupying the Bekaa Valley was pivotal for success—and the IDF was hoping that Syria wouldn't get involved. These hopes were quickly dashed when Syrian MiGs appeared along the Lebanese border, trying to engage Israeli aircraft.

Throughout the summer of 1982, Israeli fighters routinely made short work of the Syrian MiGs operating in Lebanon. On the ground, however, Syrian troops were under siege in Beirut, and the Syrian Air Force needed a reliable reconnaissance asset to help ground commanders plan their defenses accordingly. To this point, the Syrian Air Force had been using MiG-23 "Floggers" for impromptu reconnaissance missions but, just like the MiG-21s, the Floggers were easy prey for Israeli F-4s and F-15s. Thus, to maintain a watchful eye on the IDF positions surrounding Beirut, the Syrian Air Force sent its MiG-25RBs over the city. Thus, it fell to the high-flying Foxbats to obtain photographic intelligence during one of the most hazardous periods in the conflict.

Israeli F-15s had already claimed two MiG-25P kills from the previous year. But they had yet to challenge the faster MiG-25R-series reconnaissance version. A few pilots recalled catching brief glimpses of the R-variant during the latter's flyover on behalf of Egypt. But no Israeli aircraft had yet downed a MiG-25R. Throughout June and July 1982, the Israeli Air Force made several attempts to intercept the MiG-25R—all of which ended in failure. Indeed, it appeared as though the reconnaissance Foxbat was untouchable. These initial failures, however, galvanized the IDF, "which pooled resources and devised a combined operation aimed at stopping the overflights."

Flying at altitudes of 70,000 feet at speeds of Mach 2.5, the MiG-25Rs were, theoretically, safe from the Hawk SAM batteries that had tried to engage them during the days of Det 63 and Det 154. However, the Israeli Air Force devised an innovative two-tiered attack plan that combined the improved Hawk missile system with the AIM-7F "Sparrow" air-to-air missile. According to the plan, "the SAM battery would be specially-deployed on high ground in Lebanon so as to close the gap between the missile's engagement envelope that topped out at 55,000 feet and the MiG-25R's cruising altitude." Theoretically, as the MiG-25 attempted to evade the ground-based radar, it would vector itself right into the jaws of an Israeli F-15, meeting its demise from a Sparrow missile.

The ambushes were set in place but, like many innovative

techniques, the first iteration failed. On August 12, 1982, the radar detected a Foxbat on approach to Beirut. However, the Hawk battery failed to obtain radar lock on the MiG. "Subsequently, a technician manning the battery noticed that the Improved Hawk system was optimized to lock onto either 'fast and low' or 'high and slow' targets. For this reason alone, the weapon had failed to lock-up the MiG-25R. The technician duly devised a simple override for the battery's guidance system, that enabled the Improved Hawk to engage 'high and fast' targets."

With these adjusted radar parameters, the battery had its second chance on August 31, 1982. A MiG-25R was making its run over Beirut. This time, however, the recalibrated radar locked on to the MiG and launched two SAMs into its flight path. Winding their way to the target, the SAMs exploded short of the MiG, but rattled the plane enough to damage it. Trailing smoke, but otherwise undaunted, the still-airworthy MiG descended to a lower altitude—right into the engagement envelope of the F-15 piloted by Captain Shaul Schwartz. Selecting his AIM-7F missile, Schwartz watched in delight as the Sparrow liquidated the ailing MiG. Aside from netting the IDF its *third* Foxbat kill, the Israelis had an opportunity to inspect the wreckage following the engagement. Fittingly, the Israeli Air Force credited both Schwartz and the Hawk battery for the MiG kill.

Throughout the 1990s, Syrian MiG-25s regularly patrolled the Israeli border. The IDF did likewise, but there were no further shootdowns involving Israeli F-15s and Syrian Foxbats. Indeed, the MiG-25s saw little action until the outbreak of the Syrian Civil War in 2011. That conflict, growing out of the wider "Arab Spring" movement, called for the removal of President Bashar al-Assad. Starting with mass demonstrations, the discontent quickly escalated into armed conflict, becoming a multi-sided civil war involving the Assad regime (and its international allies); a loose alliance of rebel groups including the "Free Syrian Army;" Salafi jihadists; the Kurdish-Arab Syrian Democratic Forces (SDF); and the Islamic State of Iraq and Syria (ISIS), with a number of other nations directly or indirectly supporting the various sides (Iran, Russia, the United States, et al). The Syrian Civil War is ongoing as of 2020.

The first reported activity of Syrian MiG-25s in the civil war occurred on February 8, 2014, when two Turkish F-16s were scrambled to intercept a Syrian MiG-25 approaching the Turkish border. On March 27 of that year, a Syrian MiG-25 was captured on film, ostensibly flying a ground attack mission over the Hama Eastern countryside. Prior to these two engagements in 2014, no Syrian MiG-25s had been seen since the start of the civil war. Possible reasons for the Foxbats' early absence include the nature of the conflict (lending itself to close air support missions rather than high-altitude operations) or the difficulty in keeping the MiG-25s operational. Since 2018, reports abound that the Syrian Air Force currently has only one or two MiG-25s remaining, none of which are airworthy. Whatever their current status, and whatever the outcome of the civil war, there can be little doubt that the MiG-25 is in the twilight of its service in Syria.

Libya

During the reign of Muammar al-Gadaffi, the State of Libya had an adversarial relationship with the West. Curiously, one of the focal points in the ongoing strife with Libya was the Gulf of Sidra. In 1973, Gaddafi declared the entire Gulf of Sidra as Libyan territorial water. Essentially, he had drawn a line from the mouth of the Gulf of Sidra stretching from the city of Misurata on the east to Benghazi on the west, laying claim to more than 150,000 square miles of the Mediterranean

Sea. Any encroachment into the Gulf of Sidra, he added, would be seen as an act of aggression.

Most of the world, however, only recognized twelve miles from a country's shoreline as the limit for its territorial waters. Countering Libya's argument, the US referenced the 1958 Convention on the Territorial Sea and Contiguous Zone. Although Libya had not been a signator, the territorial sea convention nevertheless stated that a country could include a coastal embayment within its territorial waters only if it spanned 24 miles or less.

The Gulf of Sidra was 275 miles long.

Thus, because the Gulf of Sidra did not meet the convention's territorial standard, the United States refused to recognize what Gadaffi had declared as his "Line of Death."

Undeterred by the international rhetoric, and a seemingly-outdated 1958 treaty, the Libyan leader proudly beat his chest and bragged that he would shoot down any US or NATO aircraft that vectored into the Gulf of Sidra.

Meanwhile, Gaddafi was still trying to rebuild his forces in the wake of his disastrous war with Egypt. Among his top priorities was rebuilding the Libyan Arab Air Force (LAAF). Appointing Colonel Salleh Abullah Salleh as the new LAAF commander, Gaddafi gave him nearly unlimited latitude to revitalize Libya's air squadrons. Salleh had been trained in the US (during friendlier days) and was conversant on Western/ NATO aircraft and their attendant tactics. "Salleh attempted to solve the problems the air force was facing through initiatives in several different directions." First, he re-established ties with the Soviet Union and, after exhaustive negotiations, secured the delivery of several new aircraft. These included more than sixty MiG-25P interceptors and MiG-25R reconnaissance jets. Salleh also negotiated the purchase of several dozen Sukhoi Su-22 fighter-bombers and Tupelov Tu-22 tactical bombers.

The Libyan MiG-25PDS interceptor was purportedly the pride of Gaddafi's air fleet. Intercepts with American aircraft over the Gulf of Sidra were frequent. By the mid-2000s, most of the Libyan Foxbats had been grounded, due to lack of spare parts and other serviceability issues. At least a handful of these Libyan MiGs were restored to operational status during the Libyan Civil Wars of the 2010s. (US Navy)

The new acquisitions gave Muammar al-Gaddafi an unshakable confidence in Libya's Air Force, prompting him to reiterate his admonitions regarding the Line of Death. Normally, the US would have no cause for concern over the mindless "saber-rattling" of a Third World dictator. However, Gaddafi's rhetoric was directly addressing the navigational exercises that the US Navy (and other seagoing fleets) had practiced for decades. Indeed, maneuvering on the high seas allowed blue-water navies the opportunity to hone their skills as wartime seafarers. For naval aviators, these navigational exercises gave them the ability to test-fire missiles and engage in mock dogfights under realistic fleet conditions. Thus, Gaddafi's territorial delusions posed a direct challenge to the rights of international seagoing.

Under President Jimmy Carter, the US Navy had begun sending its battle groups into the Gulf of Sidra, both in defiance of Gaddafi's blustering and in defense of international maritime standards. The US State Department, meanwhile, announced that "oceans beyond the territorial seas are 'high seas' on which all nations enjoy freedom of navigation and overflights, including the right to engage in naval maneuvers."

In 1979, the United States launched an official Freedom of Navigation (FON) program to challenge positions such as Libya's. To this point, FON had simply been a tradition, carried out in practice by the US Navy and other maritime fleets. Going forward, however, FON would be an official program backed by authority of the US government and enforceable by the US military. It was through the official Freedom of Navigation program that the US Sixth Fleet conducted maneuvers in the Mediterranean, prompting interceptions from Libyan MiG-25Ps every time. Most of these close encounters ended without incident, and became little more than "photo ops" for the pilots of either side.

Despite the availability of open sources, the service history of the Libyan MiG-25s remains largely obscure. For years, most of the available imagery depicting Libyan Foxbats came from US Navy planes, whose pilots had taken amateur photographs from the cockpit. At times, the American pilots were so impressed by the MiG-25's handling, that they speculated these Foxbats were being flown by Soviet pilots.

Following Salleh's acquisition of the MiG-25P and MiG-25R, several of Libya's best pilots underwent conversion training. Pilot Ali Tani recalled his 1978 conversion training as such:

"After doing post-delivery test-flights, Soviet pilots never flew any Libyan MiG-25s. There were Soviet instructors in Libya, but only to help us in training and the working up of our units. They were not permitted to live on our air base, they lived in a hotel nearby, from which they could only move escorted by guards. The Soviets became involved in several attempts at smuggling liquor and gold, and often got themselves into problems with our customs. The initial plan was for the [Libyan Air Force] to establish six squadrons equipped with MiG-25s, including Nos. 1005, 1010, 1015, 1025, 1035, and 1055. Four units became operational by 1981: No. 1005 and 1025 at al-Jufra/Hun; No. 1015 at UINAB [Ukba Ibn Nafi Air Base], and No. 1035 at Mitigia. The latter unit included a reconnaissance detachment equipped with six MiG-25Rs. Stocks of spares and maintenance facilities were established at Sebha. All units had regular detachments deployed at Misurata and Benina."

During these initial fielding days, only one Foxbat is known to have crashed: a MiG-25PU trainer on November 21, 1978. According to Tani, the plane suffered an engine failure, prompting both crewmen (pilot and trainee) to eject. Sadly, one of the crewmen drowned after landing in the sea off the coast of Tripoli.

As for the US Sixth Fleet, routine encounters with the Libyan MiG-25s continued unabated. One notable interception occurred on August 18, 1981. Lieutenant Lawrence Muczynski, an F-14 pilot assigned to the US Navy's Fighter Squadron VF-41, had his first encounter with a Libyan MiG-25. Flying a routine Combat Air Patrol (CAP) from the USS *Nimitz*, Muczynski and his wingmen were engaged by several Libyan probes. Whenever these Libyan aircraft ventured too close to the carrier's training space, the F-14 Tomcats would intercept and escort them back to a proper distance. On this day, Muczynski's plane (call-signed: *Fast Eagle 107*), intercepted a Foxbat at close range. "Half of the time, we would intercept them and they would let us join right to the wing," he said. "It was pretty surreal to be a couple feet from a MiG-25 that, up to now, had only been a picture in a book." Muczynski also marveled at the massive size of the Foxbat's engines. After escorting the wayward MiG to the edge of the training area, the Libyan pilot ignited his afterburner and sped away from *Fast Eagle 107*. During these intercept maneuvers, Muczynski's wingman would follow the Libyan aircraft from a mile behind, ready to engage the bandit if it attacked Muczynski during escort. Returning to the Nimitz that evening, Muczynski compared notes with his fellow aviators. They, too, had seen their share of Libyan aircraft throughout the day—including MiG-25s and Mirage F1s. The following day, Muczynski would participate in the infamous Gulf of Sidra Incident, wherein he and his wingman, Commander Hank Kleeman, were fired upon by two Libyan Su-22s. During the ensuing dogfight, Muczynski and Kleeman each downed one of the offending Su-22s, demonstrating the effectiveness of the F-14 Tomcat and becoming national heroes in the process.

Despite losing a pair of Su-22s in the Gulf of Sidra, the LAAF seemed none too shaken. Within an hour of Muczynski's return to the *Nimitz*, the Libyans continued sending their MiGs and Sukhois into the Gulf of Sidra. Lieutenant Ed Andrews, a pilot from Fighter Squadron VF-84, who was likewise flying from the *Nimitz* that day, went airborne on another CAP, hunting for what AWACS had identified as incoming MiG-25s. His wingman for this Foxbat hunting mission was Lieutenant Junior Thomas of VF-41. From the backseat of Thomas's F-14, his Radar Intercept Officer, Lieutenant Commander Paul Williamson recalled:

"After a short time on station, we were vectored almost

due west to intercept two high-speed, relatively high-altitude aircraft which appeared to have launched from Misurata. We acquired the aircraft on radar and completed the intercept, *identifying the aircraft as Foxbats.*"

As it were, these two Foxbats were MiG-25Ps from the No. 1015 Squadron, forward-stationed at the Misurata Air Base, led by Major Khalid Maeena. Like his comrades, Maeena had orders to "chase away" any and all US aircraft in the area. Today's attempt, however, failed because the nearby US Navy electronic warfare planes had jammed his radar, thus preventing him and his wingman from locking on to the American F-14s. Although there was much tension in the air stemming from the Su-22 shootdown just hours earlier, this Foxbat intercept devolved into nothing more than a routine aerial show of force, with no exchange of fire. As Williamson explained:

"Naturally, we were excited, but the intercept and escort of the Foxbats was relatively routine. Upon intercept, they made a few mild 360 degree turns and then appeared to be returning to base. We broke off, but then had to intercept them again when they steadied up on an easterly heading (i.e. towards the fleet). Eventually, they switched on their afterburners and departed, climbing to the west. I was impressed by the size and acceleration of the Foxbat, but because of its size, weight and wing configuration, I don't believe it would present much of a threat in a conventional turning fight. Its acceleration in afterburner is impressive, but understandable in view of its published high-altitude speed."

Remaining airborne for four hours, Andrews and Thomas returned to the *Nimitz* without a kill.

Later that evening, the two downed Su-22 pilots appeared on Libyan state television, alive and well, both having survived their ejections. Major Maeena also appeared on the evening's broadcast—claiming that his MiG-25 had chased away six F-14s "that had been on their way to attack many Libyan cities." He went on to thank Gaddafi and the Soviet Union for supplying him the MiG-25, a plane that enabled him to "defend his homeland from the American aggressors, and shoot down all six of them as they tried to land and hide on their carrier ship." Of course, these were lies published for propaganda purposes. The reality had been that the Americans bested the MiG-25 and the Su-22, two planes that were purportedly the pride of Gaddafi's air fleet.

Ali Tani, whose own MiG-25P had intercepted an F-4J Phantom earlier that week, explained that:

"Essentially, our Soviet aircraft [including the MiG-25] were badly under-equipped electronically. Within one minute of receiving an initial intercept vector from our ground control, the EA-6Bs and EP-3s deployed by the US Sixth Fleet in support of their interceptors had blinded our radars and muted our communications. Our radars were useless. We could neither see what's going on ahead of us, nor ask our ground control for another vector or advice. Our ground control could not listen to American radio transmissions. If we acquired anything with our radars, they would implant decoy radar signatures into our systems, which we would pick up as 'real' bogeys. We were successively sending our interceptors after fake targets only for our pilots to find nothing but F-4s and F-14s on their tails. These were traps to bring us to their turf, while we thought we were chasing a real target…"

As the Libyan government tried to come to grips with the recent debacle over the Gulf of Sidra, there was a strong tendency to blame the Soviet Union for delivering aircraft of "inferior quality." The Soviet advisors retaliated by blaming the poor skill and motivation of LAAF personnel. In reality,

A non-flying MiG-25 sits on the tarmac at a Libyan air base in 2009. Within two years of this photograph's date, Libya would erupt into the first of its civil wars following the "Arab Spring." (Rob Schleiffert)

however, the MiG-25s of the LAAF were not much different than the domestic versions flying over the Iron Curtain. Eventually, the USSR gave Libya newer versions of the Foxbat, but given the comparative quality of Western fighters, there was only so much the MiG-25 could do.

Another notable Foxbat showdown occurred in 1986. Tensions with Libya remained high. Despite the LAAF's poor showing in the 1981 Gulf of Sidra Incident, Gaddafi had shown no signs of curtailing his rhetoric. From his support of terrorism (including the hijacking of TWA Flight 847), to his attempts at nuclear power, and his insistence on the nautical "Line of Death," Muammar al-Gaddafi was becoming a threat that the US could no longer ignore.

Tensions further escalated on January 26, 1986, when two F/A-18 Hornets on a routine CAP in the Mediterranean were met by two brand new Libyan MiG-25PDS interceptors—recently acquired from the Soviet Union.

"Vectored by an E-2C [early warning aircraft], the Hornets circumnavigated around the approaching MiGs and then took position behind the two Libyans. After being tailed by the Americans for nearly 10 minutes, both MiG-25s broke off and returned to base.

That February, during another Freedom of Navigation exercise, American F-14s and F/A-18s intercepted several dozen LAAF Foxbats over the Mediterranean. The US pilots noted that these Libyan MiG-25s were flying more aggressively than they had done in years prior. Commander Robert Stumpf, commanding officer of Squadron VFA-132, postulated:

"This aggressive Libyan response may have been due to emphasis from Soviet advisors to demonstrate existing capabilities. Also, the Libyans may have perceived a hostile intent…but felt more confident after [US Navy] forces withdrew without challenging their claim to the Gulf of Syrte [Sidra]."

The Libyan pilots, however, denied any influence from Soviet advisors. Rather, this newfound aggression was borne of a desire to "save face" from their humiliation during the 1981 shootdown and to assert their dominance over what they considered to be Libyan waters.

By and large, however, the US Navy pilots were unimpressed by their Libyan counterparts. None had demonstrated exceptional flying skills, and it seemed that even their best equipment was overmatched by the F-14 Tomcat. As Lawrence Muczynski recalled: "We had been briefed that the Libyan pilots did not like to fly at night, and didn't care much for flying over the water."

Still, the sheer speed of the Foxbat was enough to vex even the most skilled of naval aviators. Libyan Foxbat pilot Ali Thani stated:

"Our commanders were concerned about [the]

possibility of American warships crossing the Line of Death unobserved. We had no AWACS so the idea was born to use Il-76 transports to track down the USN carriers. They usually flew north of Tripoli in an east-west pattern, and were frequently intercepted by Hornets and Tomcats…In one instance, we exploited the American preoccupation with our Il-76. The crew of the transport advised us about USN warships they could track with radar. I was scrambled with my wingman [in their respective MiG-25s] and we were intercepted by a pair of Hornets. But we easily out-accelerated the Americans and passed high above one of their carriers, signaling them that they can't hide from us."

On another occasion, two MiG-25s intercepted an EA-3B flying from the USS *Coral Sea*—before any Tomcats or Hornets could arrive to ward off the Libyan bandits. The Libyan MiGs came dangerously close to the surveillance plane—passing underneath its fuselage and within range to see the detailed markings of its parent squadron. However, the MiG-25s held their fire, as the lone US Navy plane was unmistakably beyond the "Line of Death," and American fighters were only moments away.

The last notable interception of a Libyan Foxbat occurred on March 24, 1986. The day prior, two American F-14s from Squadron VF-102 "Diamondbacks" (flying from the USS *America*) crossed the imaginary "Line of Death" and were engaged by two Libyan SAMs. The missiles, fired from the air defense station near Surt, missed the Tomcats and fell harmlessly into the Mediterranean. Undeterred, the Libyans fired two more missiles, but they were quickly jammed by a nearby EA-6B Prowler. Later that day, two Tomcats from Squadron VF-33 intercepted a pair of MiG-23s. The Tomcats acquired missile lock on both aircraft, but the "Floggers" fled as soon as they realized they had been locked on by an F-14.

As the carrier group's daily duties began on the morning of March 24, two additional Tomcats had barely settled into their CAP maneuvers when they were engaged by two MiG-25PDS interceptors flying from Benina Air Base. Because the Tomcats and their parent fleet had vectored into the farcical "Line of Death," the Foxbats were scrambled with orders to shoot down any airborne intruders. Libyan pilot Ali Thani recounted the mission as such:

"I did not expect us to be successful but was determined to carry out my orders. We climbed to [19,600 feet]. The GCI vectored us to about 30 kilometers from the nearest target and then ordered me to activate my radar, acquire a target, and open fire. This was my first encounter with Tomcats and I was expecting them to be armed with Phoenix missiles. For this reason, I decided not to lock on them and fire one of my R-40RDs, but instead wanted to get close and wait for an opportunity to deploy my short-range R-60MKs. While approaching, we maneuvered as so often before: whenever they turned to one side, the GCI redirected us. This was repeated several times until we got closer and I turned directly towards one of [the] Tomcats, trying to lock-on with a missile. It did not work because the Tomcat disappeared from my view too soon."

The Tomcats, meanwhile, were preparing to maneuver around the incoming Foxbats. Under the Rules of Engagement, neither Tomcat could fire unless fired upon. The two F-14s thus dragged the MiG-25s into a descent, leveling off at nearly 5,000 feet. The MiG-25 was hardly maneuverable, but even less so at lower altitudes. Indeed, below the 10,000-foot mark, the F-14's tactical advantage increased exponentially. Outmaneuvering both MiGs, the Tomcats turned and sallied into position directly behind the two Foxbats. The pilot of the lead F-14 set his selector switch to "Guns" while his wingman reported the MiG-25s' "excessive

The Israeli F-15 Eagle, nicknamed "Baz." During the Arab-Israeli air wars of the 1980s, Israeli F-15s were the scourge of Syria's MiG-25s, regularly besting the Foxbat in aerial skirmishes over Lebanon. (US Air Force)

actions and intent," thus requesting permission to fire. The lead Foxbat (piloted by Ali Tani) slowly vectored off to the right, followed by his wingman, before both MiGs reversed course to the left. Watching the MiGs' flight pattern, the trailing Tomcat again requested permission to fire.

Minutes passed, yet there was no reply from the USS *America*.

Both Tomcats continued shadowing the MiG-25s, until both Foxbats suddenly ignited their afterburners, and disappeared over the southern horizon. By the time the air commander onboard the *America* granted permission to open fire, both MiGs were gone and the two F-14s were already on their way to the refuel point. Although no shots had been fired during this aerial skirmish, it nevertheless underscored the ongoing tensions between Libya and the US.

Following another shootdown incident over the Gulf of Sidra in January 1989 (where American F-14s downed two outdated MiG-23s), Gaddafi finally dialed down his rhetoric and made diplomatic strides to "mend fences" with the international community. Moreover, he accepted the realization that his military posed no viable threat to the West.

As it turned out, by the dawn of the new millennium, Gaddafi's days of ruling Libya were numbered. In the wake of the broader "Arab Spring" uprisings, Libya devolved into a civil war in February 2011. The eight-month conflict ended with the death of Muammar al-Gaddafi and victory for the Libyan Rebels and their "National Transitional Council." However, unresolved issues following the end of hostilities led to the *Second* Libyan Civil War, which erupted in 2014 and remains ongoing.

Throughout the second iteration of Libya's Civil War, the MiG-25 maintains a semi-active role in the Libyan Air Force. By the start of hostilities in 2011, the entire MiG-25 fleet had been grounded due to lack of funds and spare parts. These grounded MiGs had been stored and, miraculously, were spared during the NATO-led bombing campaigns in support of Libyan rebels.

After the start of the second civil war, however, Libyan forces under the New General National Congress (which controlled a number of Libyan air assets from the Gaddafi regime) began restoring the derelict MiG-25s in an effort to push them back into service. These technicians succeeded in restoring at least a few Foxbats, but one crashed on May 6, 2015 while attacking a civilian airport controlled by opposition forces. The pilot ejected but was promptly captured by the enemy, who also claimed to have downed his MiG.

Algeria

At the edge of the Mediterranean, the People's Democratic Republic of Algeria was the first recipient of the export MiG-25 in 1978. In the aftermath of their independence from France, the Algerian government wanted to build its air force into the premier aviation arm of the Mediterranean.

Maintaining ties with both the US and the Soviet Union, Algeria routinely accepted military equipment from both superpowers.

To affect their air defense capabilities, the Algerian Air Force approached the Soviet Union for delivery of the MiG-25. Algeria ultimately purchased a total of 48 MiG-25PDS interceptors, MiG-25RBV reconnaissance platforms, along with MiG-25PU and MiG-25RU trainers for the respective interceptor and recon variants. A handful of these new acquisitions were presented to the public on November 1, 1979 whilst celebrating the 25th anniversary of the Algerian Revolution. The early Algerian Foxbats initially saw service with the 120th Independent Squadron (flying the MiG-25 interceptor) and the 515th Reconnaissance Squadron (flying the R-series variants).

Although the Algerian Foxbats never saw true "combat," they were nevertheless active throughout the Mediterranean. During the 1980s, for example, Algerian MiG-25s routinely flew reconnaissance missions along the Spanish and Moroccan borders. In 1988, these same MiGs flew protective air patrols during the PLO's congress in Algeria. More recently, Algerian Foxbats have flown reconnaissance missions over lands occupied by Islamic extremists and have flown alongside various NATO members in joint exercises.

By most reports, the Algerian Air Force currently has between 10-13 operational MiG-25s. In the coming years, however, these Foxbats will likely be retired in favor of the MiG-29 and Sukhoi Su-30.

★

Chapter 4
From Babylon to the Subcontinent

Saddam Hussein, the President of Iraq, rose to power in 1968 following the Ba'ath Party revolution. As he ascended to the presidency, however, Saddam ruled Iraq with a brand of brutality reminiscent of Hitler and Stalin. Consolidating his power into a dictatorship, he seemed poised for a long, prosperous rule…until his fortunes changed in the wake of the Iranian Revolution.

Although technically allies, the Iraqis and Iranians had never fully trusted one another. Border disputes between the two nations had been ongoing for years. And since the death of Egyptian President Gamal Nasser in 1970, both Iraq and

The RB-variant of the Foxbat (pictured here, landing at Werneuchen airfield) was exported to the Iraqi Air Force for service in the latter's burgeoning war against Iran. (Rob Schleiffert)

Iran had been jockeying to become the dominant force in the Middle East. After the Shah of Iran's downfall, however, Saddam Hussein took warning of the Ayatollah Khomeini's rhetoric. Saddam, and many within his Ba'ath Party government, were Sunni Muslims. The majority of Iraq's citizens, however, were *Shiite* Muslims—just like Khomeini and his disciples. Fearing that the Ayatollah's rhetoric would galvanize Iraq's Shiite majority, the "Butcher of Baghdad" launched a preemptive invasion of Iran on September 22, 1980. Simultaneously, Saddam had hoped to take advantage of the instability following the Shah's exile, and seize key petroleum fields along the Iranian border. For the next eight years, the ensuing Iran–Iraq War would cost thousands of lives and end in a bloody stalemate.

Like his counterparts in Libya and Syria, Saddam had garnered close ties with the Soviet Union. In 1971, he signed an official Treaty of Friendship with the USSR, granting him access to the latest in Soviet fighters and weaponry. Throughout the 1970s, Saddam steadily grew the size of the Iraqi Air Force, equipping it with modern fighters such as the MiG-23, Su-22, and eventually the MiG-25 Foxbat.

Saddam was already familiar with the Foxbat, having seen it during its high-speed flyovers of Iran, Iraq, Israel, and Turkey. In fact, during the reign of the Shah, Iranian F-4 Phantoms regularly tried to intercept the MiG-25 during its many incursions into Iranian airspace. Not coincidentally, however, the Soviet Air Force discontinued its Iranian flyovers after the Shah acquired the F-14 Tomcat. Seeing what the MiG25P and MiG25R could do, Saddam solicited the purchase of both variants for the Iraqi Air Force.

By the early 1980s, Saddam's air force had received 19 MiG-25PD and PDS variants, along with nine MiG-25RBs, and seven MiG-25PU trainers. The MiG-25PD and PDS versions were concentrated in the No. 96 Squadron at al-Taqaddum Air Base. Some sources, however, indicate that

The Iranian F-14 Tomcat. Exported to Iran from the United States during the reign of Shah Pahlavi, the F-14 was the scourge of many Iraqi Foxbats. Like its counterpart F-15 Eagle in Israeli service, the Iranian Tomcats proved superior to the MiG-25 and other Soviet-built airframes in Arab service. (Shahram Sharifi)

these MiG-25 interceptors belonged to the No. 1 Fighter/ Reconnaissance Squadron—purportedly staffed by the best pilots in the Iraqi Air Force, supplemented by Soviet attaches. The MiG-25RBs, on the other hand, belonged to the No. 17 Fighter/Reconnaissance Squadron. By 1986, these RB-variant Foxbats were upgraded to the newer RBT standard. Throughout the course of the war, the Iraqi Air Force activated three additional MiG-25 squadrons—No. 84; No. 96; and No. 97.

During the opening volleys of the conflict, the Islamic Republic of Iran Air Force (IRIAF) scrambled every available asset, taking flight to intercept the Iraqi MiGs and Sukhois. Even before the official start of the war, Iranian F-14s had intercepted a number of Iraqi air patrols probing the Iranian border. For example, on September 7, 1980, five Iraqi Mi-25 "Hind" helicopters ventured into Iranian airspace in the Zain

An American F-15 Eagle sits in a revetment in Saudi Arabia during the lead-up to Operation Desert Storm. Four Patriot surface-to-air missile launchers are visible in the background. During the Gulf War, American F-15s from the 58th Fighter Squadron downed two Iraqi MiG-25s. (US Air Force)

al-Qaws region, attacking Iranian Army outposts along the border. Two F-14s scrambled to intercept, skillfully downing both helicopters and scoring the first air-to-air victory for an F-14 Tomcat.

Assessing the true performance of the MiG-25 during the Iran-Iraq War, however, remains a daunting task. Both sides inflated their numbers (and exaggerated their accomplishments) for propaganda purposes. The reports and descriptions of their aerial victories vary from source to source. According to most contemporary sources, however, the Iraqi Air Force achieved its first Foxbat victory on May 3, 1981. Piloted by Captain Mohammed Rayyan, the MiG-25 was scrambled to intercept an unknown aircraft that vectored away from the international air corridor near the Iranian-Turkish border. Rayyan, directed by a local radar station, intercepted the "bogey," which turned out to be an Algerian

Gulfstream III. With orders to shoot down any non-Iraqi aircraft in the vicinity, Rayyan acquired missile lock, downing the Gulfstream with an R-60 missile.

Most sources indicate that Mohammed Rayyan was the top Iraqi ace of the war, scoring between 5-10 air combat kills—making him the most successful MiG-25 pilot in history. Most of these purported victories were against F-4 Phantoms and F-5 Tigers, planes that Iran had acquired from the US during friendlier times. Rayyan ultimately met his demise in 1986 when his Foxbat was downed by an Iranian F-14.

Some reports, however, suggest that Mohammed Rayyan never truly existed—that he was simply a mythical propaganda figure created to lift morale (similar to "Uncle Sam" or "Willie and Joe"). These same reports indicate that the Iraqi Air Force never produced an "ace" pilot, neither during the Iran-Iraq War, nor during the subsequent Gulf War. Claims of multiple kills from Iraqi pilots were often reduced *drastically* once the official confirmations took place. For example, an Iraqi Mirage F1 pilot once claimed a dozen kills against the IRIAF, but upon re-evaluation from the Iraqi Air Force, his official tally was reduced to only two kills. The lack of biographical data is another indication that Rayyan may not have existed. Most sources cite 1986 as his year of death, but provide neither a specific date nor location. Iraqi and Iranian records confirm that a MiG-25RB was shot down by an F-14 Tomcat on February 15, 1986. However, this could not have been Rayyan's plane as he only flew the interceptor variants and his last purported kill was registered in June 1986.

But whether Mohammed Rayyan was a real person or simply a hero of modern folklore, his accomplishments in the MiG-25 suggest that the Foxbat was an indelible asset to the Iraqi Air Force. Other aerial victories attributed to Rayyan include the March 1985 downing of an F-4 Phantom piloted by Major Hossein Khalatbari, one of the most respected pilots in the IRIAF. To this point, Khalatbari had been somewhat of a legend within the Iranian military. As a flight commander, he led numerous air interdictions against Iraqi targets. By 1985, Khalatbari had been credited with destroying more than 20 Iraqi naval vessels, and leading the airstrike on the Kirkuk refinery. On the night of March 21, his F-4 was scrambled from Hamedan airbase to intercept a flight of incoming Iraqi jets. During the engagement, Khalatbari successfully downed a MiG-23 before his own plane was destroyed by an R-40 missile from Rayyan's MiG-25. The loss of Khalatbari was a tremendous blow to the IRIAF.

Although Rayyan's purported victories aboard the MiG-25 were impressive, the IRIAF crates could certainly hold their own against the Foxbat—especially the F-14 Tomcat. The first known engagement between Iraqi MiG-25s and Iranian F-14s occurred on May 15, 1981. The IRIAF hastily deployed a squadron of F-14s to Tactical Fighter Base 4 (TFB 4) at Vahdati, near the Iraqi border. From the opening days of the war, TBF 4 had been a frequent target for Iraqi artillerymen and attack aircraft. In an effort to re-establish air superiority over the region, the IRIAF hoped that the new-fangled Tomcats would be enough to silence the Iraqi bombardment.

Major Ali, an F-14 pilot assigned to the 82d Tactical Fighter Squadron, vividly recalls the actions of his comrades the day they arrived at TFB 4:

"Only two hours after their arrival, four F-14s and two F-4Es established a CAP west of the airfield. Within minutes, they detected six MiG-23BNs covered by four MiG-21s. We attacked and two MiG-21s were shot down, both by Sidewinders. A few minutes later, the lead F-14A RIO detected a MiG-25RB closing at high speed, but still inside Iraqi airspace. The [F-14] pilot immediately turned to attack, and within seconds a single AIM-54A [Phoenix] was launched at the target, which was still [67 miles] away."

The remains of an Iraqi MiG-25 after being destroyed on the ground by a 2,000-pound laser-guided bomb. During Desert Storm, much of the Iraqi Air Force was destroyed before it ever went airborne. Several more Iraqi planes were destroyed during the ensuing years as the US and its NATO partners enforced the Iraqi No-Fly Zones. (US Department of Defense)

The Foxbat pilot, however, detected the missile via his radar warning receiver. The Iraqi MiG-25 made a tight turn and thundered away at a speed of Mach 2.2, activating his onboard radar jammer as he fled. Major Ali commented that the MiG-25's evasive maneuvering and electronic countermeasures were marginally effective. "The Foxbat moved towards the edge of the Phoenix envelope," he said, "but the missile had a built-in, home-on jam capability and the weapon passed close by its target, exploding behind the jet. That MiG pilot was lucky. Most of the shrapnel missed, but his aeroplane was still damaged and he had to make an emergency landing at Shoaibah." By all measures, this first encounter between the MiG-25 and F-14 ended in a draw. There would, however, be several more showdowns between the Foxbat and the F-14.

By the fall of 1981, it appeared that the IRIAF had gained the upper hand along the Iraqi border. Indeed, by then, the Iraqi Air Force could only muster about 140 combat-ready aircraft. "The Iraqis now had little choice but to increase the use of their MiG-25s," said historian Tom Cooper. "It was the only type superior to Iranian interceptors in at least one aspect—speed." Meanwhile, the geopolitical irony of the situation was not lost on observers from the Iron Curtain; the Soviets watched in dismay as their Foxbat succumbed to the likes of the F-14.

Still, the MiG-25 was the fastest jet the Iraqis could muster and (aside from the Mirage F1), it was the only Iraqi jet that could stand against the newer American-built airframes used by the IRIAF. Thus, to affect the viability of their MiG-25 forces, the Iraqi Air Force intensified the training and selection for their Foxbat crews. The pilots were screened not only according to their intelligence, but to their loyalty to Saddam and the Ba'athist party line. Many of the new Foxbat jockeys had been trained by the Royal Air Force in Britain during their earlier exchange programs. The selectees then received intense conversion training on the MiG-25 from Soviet instructors.

After the initial showdown between Iranian F-14s and Iraqi MiG-25s, the Foxbats re-appeared over the battlefront in October 1981. Initially, they flew reconnaissance missions over the strategically-critical Khark Island, before initiating their bomb runs. However, these early missions aboard the MiG-25RBs were hardly encouraging for the Iraqi Air Force. Indeed, the embedded Soviet technicians were having trouble calibrating the MiG's navigational-attack system. This in turn hampered the pilot's ability to accurately engage his anticipated targets. Faulty avionics aside, the MiG-25RB missions over Khark were short-lived. As it turned out, the IRIAF intensified its Tomcat patrols over the area, thus foiling the Foxbat's mission profile.

Throughout the spring of 1982, the IRIAF tracked at least a dozen MiG-25s along the Iranian front. Typically, the Iraqi Foxbats would fly at altitudes of 62,000 feet or higher, and at speeds exceeding Mach 2. The IRIAF launched its interceptors, but more often than not, they simply couldn't catch the MiG-25s. On March 19, 1982, a flight of Iranian F-4 Phantoms was intercepted by a lone Iraqi Foxbat. One of the F-4s sustained heavy damage from the MiG-25's missile, but remained in the air, and the pilot was able to return to base. Although still vulnerable to the likes of the F-14, the MiG-25s were now penetrating deeper into Iranian airspace. Some accounts recall Iraqi MiG-25 interceptors flying within range of Tehran, battling F-4E Phantoms on the outskirts of the city. A few Iraqi pilots, however, dispute this claim, stating that the MiG-25PD and PDS were dependent on ground control radars, and that none of the Iraqi radars could reach as far as Tehran. The MiG-25RBs, however, had no such tether.

That August, the RB Foxbats returned to Khark Island. Flying at high speeds and high altitudes, these MiG-25s were difficult to catch. Even for the best F-14 crews, intercepting a MiG-25RB was a laborious task that required a painstaking combination of high-speed flying and hyper-dexterity with the avionics. The following month, as the Foxbat patrols intensified, the IRIAF began circulating their F-14s onto 24-hour CAP stations. Because the MiG-25RBs typically operated at higher altitudes, "it took the IRIAF some time to learn how to intercept them—mainly by changing patrol altitudes, positions, and speeds." On occasion, Iranian F-14s also acted as "airborne controllers," acquiring a Foxbat on its radar and then guiding other nearby aircraft to intercept it.

On September 16, 1982, two F-14s flying near Khark were advised of an incoming aircraft, travelling at 70,000 feet approaching speeds of Mach 3. Given these parameters, the F-14 pilots knew that this bogey had to be a MiG-25. Both Tomcats turned into the direction of the oncoming threat and, within minutes, they had positively identified the bandit as a MiG-25RB. Using the F-14's onboard AWG-9 radar, the RIO of the lead Tomcat locked on to the MiG and fired his AIM-54 Phoenix missile from a distance of more than 60 miles. Moments later, the Phoenix missile slammed into the offending MiG, erupting in a giant ball of fire. The pilot was said to have ejected into the Persian Gulf, but the Iranian search-and-rescue helicopters could find neither his body nor his parachute. In the shark-infested waters of the Persian Gulf, downed aviators never lasted long.

According to Iranian sources, the encounter of September 16 was the first "confirmed kill" of an Iraqi Foxbat by an Iranian F-14. An Iraqi defector, however, claimed that another MiG-25 had been lost prior to this date. This same defector explained that the Iraqi Air Force had lost nearly 100 fighters to Iranian F-4 Phantoms and F-14 Tomcats. His claims have never been confirmed. What was confirmed, however, was that the AIM-54 could catch a supersonic Foxbat in flight. Indeed, the events of September 16 were as galvanizing to the IRIAF as they were disheartening to the Iraqis.

One week later, on September 22 (the two-year anniversary of Iraq's invasion), another MiG-25RB was spotted over Tehran. Naturally, the IRIAF could not passively allow enemy reconnaissance flights over the nation's capital. Thus, the F-14s stationed at Mehrabad received priority supplies of AIM-54 missiles—the one munition proven to down a Foxbat in supersonic flight.

The next MiG-25 encounter took place near Khark Island on December 1, 1982. An F-14 Tomcat piloted by Major Shahram Rostami was flying top cover for a convoy of merchant ships near the Iranian coast. During his aerial refuel, he was alerted by ground control of an incoming bogey travelling at 70,000 feet and Mach 2.3—no doubt a MiG-25. Rostami accelerated to meet the bandit. Despite the Foxbat having preemptively jammed the F-14's radar, Rostami's RIO was able to lock on to the bandit and fire a Phoenix missile in a "snap-up" engagement while climbing to 40,000 feet. After watching the missile escape from the fuselage, Rostami banked his F-14 to the west and reduced his airspeed, so as not to merge with the incoming MiG too quickly. Maintaining his distance within the radar envelope, Rostami saw the "time-to-impact" counter on his weapons panel reach "0"— indicating that the missile should have hit the target. Just then the "hit symbol" illuminated on the radar screen, and ground control confirmed that the MiG-25 had disappeared from their scopes. Indeed, the MiG-25RB had crashed into the Persian Gulf, but the pilot was never seen again.

Determined to avenge the loss of this latest Foxbat, the Iraqis sent two MiG-25PDs into northern Iran on December 4. While searching for an opportunistic target, the two MiGs separated, unbeknownst to either pilot, two IRIAF Tomcats were in the vicinity. As soon as the lead F-14 activated its radar, the MiG-25's warning indicator went off. By now, however, the Tomcat had already fired its AIM-54 missile. Alerted to the incoming missile, the MiG-25PD accelerated to full afterburner. The Tomcat crew, meanwhile, watched as the Foxbat desperately tried to outturn the AIM-54 missile. The Phoenix would have made impact, but it malfunctioned in mid-flight, while the Foxbat continued its hasty retreat. Undaunted, however, the F-14 pilot accelerated to Mach 2.2, determined to run down the elusive Foxbat. After evading the first Phoenix missile, the MiG pilot had slowed down, confident that the F-14 was out of range and no longer a threat. The Iraqi's complacency, however, was his undoing. Within minutes, the F-14 had closed the distance and fired a second AIM-54 on the unsuspecting Foxbat. Unable to react in time, the MiG-25 was blown from the sky.

On August 6, 1983, two MiG-25PD interceptors violated Turkish airspace as a shortcut into Tabriz. On the Iranian side of the border, however, an IRIAF Tomcat was already waiting

US forces unearth an Iraqi MiG–25 at the Al–Taqqadum Air Base in 2003. Days prior to the US-led invasion of Iraq, Saddam Hussein ordered the Iraqi Air Force to bury many of its jets, ostensibly to save them from American firepower. (US Department of Defense)

Fully exhumed, the Iraqi MiG-25 is towed away from its hiding place and prepared for shipment to the United States. (US Department of Defense)

for them. Once the MiGs were within the engagement envelope for the Phoenix missile, the Iranian crew promptly fired the AIM-54. Just like the previous Foxbat interceptions, these MiG-25s began evasive maneuvers as soon as their radar sounded the alarm. The Phoenix missile, however, detonated close enough to the Foxbat duo to pepper the trailing MiG's engines and tailfin. The Foxbat was wounded, but still airworthy. His wingman, however, had left him behind, desperately accelerating back towards Iraqi airspace. Meanwhile, a nearby flight of Iranian F-5E Tigers was on its way into Iraqi airspace, delivering a strike package against critical targets in the northern provinces. IRIAF Captain Kazem Zarif-Khadem, piloting the lead F-5, took note of the ailing MiG that suddenly crossed into his flight path. Zarif-Khadem immediately thrust his F-5 to full afterburner, chasing down the Foxbat and acquiring missile lock at the "deep six" position. Two AIM-9 missiles erupted from the wings of the F-5, each meeting their mark on the MiG-25. Miraculously, the shaken Foxbat pilot ejected, and was later recovered by Iraqi troops.

Not all encounters with the IRIAF were one-sided. Iraqi MiG-25s did manage to shoot down several Iranian planes and helicopters. On many occasions, the Iraqis also used the MiG-25PD as an "aerial scout," scanning the airspace ahead of an incoming CAP or strike package. According to Iraqi sources, these vanguard MiG-25PDs would fearlessly turn to meet their pursuers, disregarding that their Foxbats were outclassed in terms of maneuverability. It was under these conditions when, on June 3, 1984, a MiG-25 interceptor shot down an Iranian F-5E Tiger over Tabriz. There was also report of an Iraqi MiG-25PD shooting down an Iranian C-130 cargo plane, purportedly carrying weapons into Iran. Yet Iranian planes weren't the only ones targeted by Iraqi MiG-25s. Breaking ranks with its Arab neighbor, Syria supported

Iran throughout the conflict, conducting aerial reconnaissance for the Iranian military. The full extent of Syria's involvement remains obscure, but on October 2, 1986, an Iraqi MiG-25PD shot down a Syrian MiG-21R.

By the dawn of 1985, the Iraqi MiG-25 bombers were having more success in their campaigns over Khark Island. The size of the island itself (1.5 square miles) made it an optimal target for the MiG-25RB's guidance system. Its strategically-rich oil refineries and storage facilities, however, meant that the IRIAF would fight to the death defending it. Still, the MiG-25 bombers were at least marginally successful in disrupting operations on Khark, destroying at various times oil terminals and tankers. At one point during their cyclic raids of Iranian oil field and air bases, the Iraqis attempted using modified French-built bombs for the MiG-25RB. The detonators functioned nicely, but the French bombs themselves were notoriously inaccurate, thus ending their use in the Iraqi Air Force. In March 1985, following an Iranian missile strike on Baghdad, the Iraqi Air Force relocated four MiG-25RBs to Kirkuk to facilitate long-range bombing raids. For one month, these MiG-25s ran daily missions, carrying four 1,100-lb bombs in raids against Tehran, Tabriz, Isfahan, and Qum.

As the war dragged on, the Soviet Air Force saw an opportunity to test its latest MiG-25BM in the skies over Iran. The BM was a specially-modified "Wild Weasel" version of the Foxbat—meaning that the aircraft was designed to suppress enemy air defenses. A handful of MiG-25BMs deployed to Iraq's H-3 Air Base in 1986. After a few weeks, however, the Soviet detachment suspended its operations after one of their MiG-25BMs was shot down by an Iranian F-14. In November 1987, the Soviets tried again—sending four new MiG-25BMs to the H-3 Air Base along with 130 technicians and a full complement of maintenance assets. This new contingent of "Wild Weasel" MiGs were armed with

AS-11 and AS-12 anti-radiation missiles. The purpose behind this new deployment was to test the BM's electronic countermeasures against the F-14 and to use the anti-radiation missiles against Iranian SAM batteries. The MiG-25BM contingent flew at least three missions over Iran. The first two sorties were successful, with the BM Foxbat flying at 68,900 feet with impunity and disabling at least one Iranian radar site. The third mission, however, was a dismal failure. Upon crossing into Iranian airspace, the MiG-25BM was engaged by a Phoenix missile from a nearby F-14. Even though the MiG pilot had initiated his electronic countermeasures—jamming the Tomcat's radar—the F-14 crew was nevertheless able to guide the missile onto the Foxbat. Luckily for the MiG, however, the Phoenix failed to detonate upon impact. The kinetic force of the missile, however, was enough to damage the plane, clipping the Foxbat's tailfin. The Soviet pilot had no choice but to abort the mission, and made an emergency crash-landing at the nearest Iraqi airbase. "To the embarrassment of the Soviet government, and its air force, and in the full view of reconnaissance satellites, the wrecked Foxbat was loaded into an Il-76 transport and flown to the USSR." Four days later, all Soviet personnel and equipment departed Iraq. One final detachment of MiG-25BMs returned in July 1988, towards the end of the conflict. Their mission was to validate the effectiveness of two upgraded anti-radiation missiles against the Iranian ADS-4 Early Warning Radars. Of these latter-day MiG-25BMs, one is known to have destroyed a radar site near Hamedan.

By 1988, however, both Iran and Iraq were eager to end the conflict. After eight years of conflict, neither side had made any significant gains against the other. That summer, the United Nations negotiated a peace settlement between the warring parties. Under the banner of UN Resolution 598,

The Iraqi MiG-25 is now on display at the National Museum of the United States Air Force. When it was exhumed by American ground forces in Iraq, the wings were missing. Still, this Iraqi specimen remains the most complete example of a MiG-25 currently in American custody. (US Department of Defense)

Saddam Hussein and the Ayatollah Khomeini accepted the terms of the ceasefire.

The Iran-Iraq War thus ended on August 20, 1988.

Although both sides claimed victory, the Iraqis and Iranians had essentially fought each other to a draw. Saddam had failed to annex the Iranian territories he desired, and Khomeini had failed to topple Iraq's regime or decisively defeat its military.

Assessing the MiG-25's performance in air-to-air combat depends largely on the origin of the source. Iraqi, Iranian, and Western sources often contradict one another. Most records, however, confirm that Iraqi MiG-25s achieved between 10-19 aerial victories. The Iraqis claim to have lost only four MiG-25s, but other sources have claimed their losses as high as 10. Post-bellum analyses aside, there can be little argument that the MiG-25 proved its resiliency, even if it did fall victim to the IRIAF's fighter fleet.

Desert Storm

Aside from the untold cost in human suffering, the Iran-Iraq War had left Saddam Hussein straddled with a multi-billion-

Another MiG-25 stands partially destroyed and vandalized at the Al-Asad Air Base in Iraq, following the US-led invasion of 2003. As Iraq established its post-Ba'athist government and military, the remaining MiG-25s were scrapped along with most of Iraq's latter-day Soviet equipment. (US Marine Corps)

dollar war debt—most of which had been financed by Kuwait. Prior to the war, Iraq had almost no foreign debt and more than $35 billion in cash reserves. By 1989, however, Iraq had spent nearly $60 billion in arms purchases alone.

But rather than pay his debt to the Kuwaiti government, the "Butcher of Baghdad" simply invaded his neighbor to the south. To justify the invasion, Saddam reignited the long-standing border dispute between the two countries. He also made false allegations that the Kuwaitis had been slant-drilling Iraqi oil and that they were deliberately trying to keep the price of oil low by producing beyond OPEC quotas. Kuwait held ten percent of the world's oil reserves and generated 97 billion barrels of crude each year. Thus, Saddam reasoned that if he could not repay his debt, he would simply annex the tiny emirate and take over its petroleum industry.

Thus, on the morning of August 2, 1990, more than 100,000 Iraqi troops and several hundred Iraqi tanks stormed across the border, the spearhead of an eighty-mile blitzkrieg into Kuwait City. Encountering only piecemeal resistance, Iraqi tanks thundered into the heart of the Kuwaiti capital, assaulting the city's central bank and carrying off with its wealth.

The invasion drew fierce condemnation from the international community and prompted the United Nations to demand Saddam's withdrawal. Undeterred by the rhetoric, the Iraqi dictator massed his forces along the Saudi Arabian border and dared the world to stop him. He was certain that his army—the fourth-largest in the world and equipped with the latest in Soviet weapons—would make short order of any rescue force that came to liberate Kuwait. He wagered that the Americans would lead a military response against Iraq but, as he famously quipped, America was "a society that cannot accept 10,000 dead in one battle." He was confident that after the Americans had suffered a few thousand casualties, they would sue for peace on Iraq's terms.

Economic and military sanctions soon followed while President George HW Bush authorized the first US deployments to the region. Within days, the aircraft carriers *Saratoga*, *Independence*, and *Eisenhower* were steaming towards the Persian Gulf while coalition air squadrons poured into Saudi Arabia by the hundreds. The first wave of deployments became known as "Operation Desert Shield"—a deterrent against Saddam Hussein lest he try to invade the Kingdom of Saud.

But for as tough as Saddam's army sounded, his air force was primitive by NATO standards. Much of the Iraqi Air Force's vitality had been eroded by the Iran-Iraq War. Still, it was the largest air force in the Middle East, with some 934 combat-capable aircraft in its arsenal, including the MiG-25.

Theoretically, the Iraqi Air Force should have been "battle-hardened" by their conflict with Iran, but Saddam's postwar purges had stripped the Iraqi Air Force of its best leaders. In the wake of Saddam's brutal crackdown, training and readiness within the Iraqi Air Force quickly ground to a halt.

In November 1990, as coalition forces poured into Saudi Arabia, the UN passed Resolution 678. The resolution, for what it was worth, gave Saddam Hussein a deadline of January 15 to withdraw his forces, or face military action. Still, the Iraqi dictator showed no signs of backing down. Thus, it came as no surprise when, on the morning of January 15, 1991, coalition forces awoke to the news that Saddam Hussein had reached his deadline—and had made no effort to withdraw from Kuwait.

The next day, President Bush announced the start of the military campaign to eject the Iraqis from the war-torn emirate. Operation Desert Shield had just become Operation *Desert Storm*. On January 17, at 2:38 AM, Baghdad time, the first wave of the coalition's air campaign destroyed Iraqi radar sites near the Saudi border. For the next five weeks, coalition air forces pounded away at key targets within Iraq and Kuwait.

Saddam, for his part, had not been idle during the weeks leading up to Desert Storm. He had prepared his fighter and interceptor squadrons at air bases including Al-Taqqadum, Al-Asad, Mudaysis, H-2, and H-3. Unlike the Iraqi Army, whose frontline formations were manned mostly by conscripts, the Iraqi Air Force had the best-trained and highest-quality personnel. Saddam confidently instructed his airmen to destroy the "pigs" who were about the descend onto Iraq. On that first night of the air war, the MiG-25 scored its first and only kill of an American aircraft—downing a US Navy F/A-18 Hornet piloted by Lieutenant Commander Scott Speicher. It was also the first time an American F/A-18 was lost to enemy fire.

Shortly after 3:00 AM on January 17, US naval air squadrons from the USS *Kennedy* and *Saratoga* sent a combined "strike package" into Iraq. The aerial task force's mission was to suppress enemy air defenses and destroy key facilities at Tammuz Air Base. Leading this strike package into Iraq were ten F/A-18 Hornets from Squadrons VFA-81 and VFA-83, flying from the *Saratoga*. As the lead element in the strike force, the Hornets crossed the Saudi border arrayed in an echelon right (i.e. "wall") formation. Each Hornet maintained a horizontal distance of 2-5 miles from another, while keeping a "stacked" vertical distance of 1,000 feet between each plane.

The F/A-18s swept ahead of the main strike force (twelve Grumman A-6 Intruders), providing front cover and suppression of the enemy's forward air defense. Once within range of Tammuz, the ten Hornets arrayed themselves into a fan formation, preparing to launch their anti-radiation missiles at the designated targets. Following behind were eight of the twelve A-6 Intruders, each of which initiated its bombing run from 25,000 feet, delivering their Mk84 2,000-lb bombs. The remaining four A-6s followed with a volley of GBU-10 Laser-Guided Bombs, destroying two prominent MiG-29 shelters.

Although this American strike package had met its first target, the Iraqi radar stations had nonetheless detected their approach. When the A-6s and F/A-18s initially entered Iraqi airspace, the only Iraqi fighters in the sky were MiG-29s, all of whom were preoccupied chasing a flight of American B-52s. When the Tammuz radar station determined that the F/A-18s were headed north to Qadessiya Air Base, the No. 84 Squadron scrambled a MiG-25 to intercept.

Taking flight to intercept the American F/A-18s was Lieutenant Zuhair Dawoud—one of four Foxbat pilots on the "standby alert" at Qadessiya that night. As Dawoud recalled:

"The Air Defense Telephone rang and I answered. There was a guy screaming at the other end of the line: 'MiG-25 IMMEDIATE TAKE-OFF!' So I hurried to the aircraft. In fact, the technicians were ready for this moment, as was the jet. So, takeoff was exceptionally fast—I was airborne just three minutes after I had received the call. After takeoff, I switched to safe [secure] frequency and established contact with GCI [ground control intercept] of the Air Defense Sector. The sky was clear with very good visibility. The GCI started to give me to a group of aircraft that had penetrated Iraqi airspace to the south of the base."

Upon takeoff, Dawoud climbed to 26,000 feet, accelerating to Mach 1.4, and set the Smerch-A2 Radar to "Standby" as the system warmed up. Several miles in front of him, beyond the darkness, was the American flight leader, Commander Michael T. Anderson. The flight leader's Hornet sat at the apex of the fan formation. From a range of about 70 miles, the MiG-25 populated on Anderson's radar. "I immediately knew it was an enemy airplane," he said. "I could see the afterburner flame, and it was an extremely long yellow flame that I had seen before on a MiG-25. As soon as I took a radar lock on him, he turned right, and at that point, he started to go around me in a counter-clockwise direction. I did a couple of circles with him." Dawoud, meanwhile, was trying to stay away from Anderson long enough to let the Foxbat's radar warm up. As Dawoud recalled: "I was 70km [48.6 miles] from the target formation when an enemy aircraft locked on to me with radar. So I performed a hard maneuver and broke the lock."

Despite Anderson's positive identification, he held his fire until the nearby AWACS could confirm whether the bogey was truly an enemy plane. However, Dawoud's Foxbat was quickly flying to the edge of the AWACS's radar scope. Without an electronic signature (because Dawoud's radar was

The Soviet MiG-25RBS served during the inaugural phases of the Soviet-Afghan War, flying reconnaissance missions in support of the 40th Army's operations. (Alex Beltyukov)

not yet transmitting), the AWACS could not confirm whether the target was hostile. After making another circle with Anderson, the MiG-25 rolled out and decelerated its afterburner which, in the dark, caused Anderson to lose sight of it. Dawoud's Foxbat then departed the scene, flying directly east and right overhead of Anderson's wingman.

Flying at the tail end of the Hornets' echelon was Scott

Speicher. As he approached the designated launch point, Speicher disengaged the autopilot and descended to 27,872 feet, preparing to launch his anti-radiation missiles.

Little did he know, however, that Dawoud's MiG-25 was about to lock him on.

Dawoud continued: "I reported what happened to the GCI and they told me to return to my original intercept course as I had 'targets at 38 km [20.5 miles]. Meanwhile, my radar became ready." Dawoud locked on to the Hornet and, from a distance of 15 miles, fired his R-40 missile. "I kept the target locked with radar till I witnessed a huge explosion in front of me. I kept looking for the aircraft going down spirally to the ground, with fire engulfing it."

As it turned out, the R-40 detonated just short of Speicher's plane, below the left side of the cockpit. The blast from the 154-lb warhead, however, flung the F/A-18 nearly 60 degrees to its right, causing a 6G turn that sheared off the plane's fuel tanks and one of Speicher's unfired missiles. The ailing Hornet plummeted into the desert floor, crashing about 40 miles south of Qadessiya. Although a few of Speicher's wingmen had seen his descent, none were certain if he had been killed or had ejected. The following day, Speicher was officially recorded "Missing in Action." Three months after the end of Desert Storm, however, Speicher's status was changed to "Killed in Action/Body Not Recovered." For the next several years, his fate remained a mystery, and the whereabouts of his remains were unknown. Eventually, the Iraqi government permitted recovery of the F/A-18 wreckage, but claimed to know nothing of Speicher's status or his whereabouts.

In December 1993, Qatari officials discovered the wreckage of Speicher's aircraft. From this, it was determined that Speicher had ejected from the plane and may have been alive for several hours after his ejection. In April 1994, satellite photography revealed human-made symbols on the desert floor near the Hornet's crash site—quite possibly Speicher's "Escape and Evade" sign. A year later, the International Red Cross sponsored an excavation of the crash site, but found no conclusive evidence regarding Speicher's fate or his whereabouts.

For several years thereafter, the US Navy officially maintained that Speicher had been downed by an Iraqi SAM. Yet, even throughout the lifespan of this "official story," many of Speicher's comrades openly doubted it. One pilot, who flew alongside Speicher during the same mission, publicly stated: "I'm telling you right now, don't believe what you're being told. It was that MiG that shot [Speicher] down." Rumors abounded that Speicher was still alive and being held prisoner in Iraq. Following the US-invasion of Iraq in 2003, however, American authorities determined that Speicher had never been captured by the Iraqi government. A break in the case finally arrived in 2009. Acting on a tip from a local Iraqi in the Anbar Province, a team of US Navy and Marine personnel unearthed the remains of Scott Speicher. According to the informant, Speicher had been buried by local Bedouins who had discovered his body. After confirming that the remains were indeed Speicher's, he was given a proper burial with full military honors at Arlington National Cemetery.

Two days after Scott Speicher's tragic death, US Air Force Captains Larry Pitts and Rick Tollini partially avenged his death by downing two other MiG-25s. Both men were assigned to the 58th Fighter Squadron, which earned the distinction of downing more Iraqi aircraft than any other unit during the conflict.

On January 19, 1991, Tollini and Pitts went aloft with a four-plane formation flying CAP into Iraqi airspace. By this point in the air war, four of Pitts' and Tollini's comrades in the 58th had scored aerial kills—downing MiG-29s and Mirage F1 fighters. Throughout Desert Shield, Pitts recalled seeing a few MiG-25s tracing the Saudi border, trying to gather intelligence on the coalition's buildup in Saudi Arabia.

"The 19th was day three of the war," said Pitts, "and the missions just kept coming." Their mission on January 19 was to "cover a large strike going into western Iraq consisting of F-16s, F-111s, and the [Wild] Weasels." Taking to the sky with Tollini and their comrades, Pitts occupied his position in the four-plane formation. Soon, the AWACS alerted them to a group of bogeys flying in the vicinity of another nearby coalition strike package.

Answering the call, Pitts, Tollini, and their wingmen pressed northeast and, within minutes, their radars detected two incoming bogeys. Ramping down to about 10,000 feet and closing the distance fast, Pitts started running "sample" radar locks onto the bogeys, seeing how they might react. Just as he had anticipated, every time Pitts engaged his radar lock, "they would break the lock, so it was obvious they were aware of us, and probably had good radar warning gear. As Pitts and Tollini closed in on the group, the Bogeys descended even lower—to 3,000 feet—"and heading due south of us in a three to five-mile lead trail formation." Soon, the bogeys descended even lower—to 500 feet (unusual considering that the MiG-25 performed better at higher altitudes). Diving deeply to meet the elusive bandit, Pitts broke through the clouds and visually identified the plane as a MiG-25.

Setting his radar lock on the MiG, Pitts knew that from this distance and altitude, it would be harder for the Foxbat to shake him. Still, the Iraqi pilot tried breaking to the right. "Not that the turning radius of a Foxbat doing 700 knots is very impressive," said Pitts, "but he tried." Staying inside the MiG-25's turning circle, Pitts got a good tone for his AIM-9 Sidewinder. Pitts fired the AIM-9, but the MiG deployed its flares, which threw the missile off its tracking. Unphased, Pitts thumbed his selector switch to the AIM-7 Sparrow and,

While the Soviets continued their war in Afghanistan, they maintained their heavy footprint in Eastern Europe, most notably in East Germany. Pictured here is a MiG-25RB stationed with one of the last Foxbat regiments in Welzow, East Germany. (Rob Schleiffert)

despite getting good tone, the missile sailed right over the Foxbat's canopy. "He continues his turn and rolls out heading north, now doing about 500 knots."

Switching back to his AIM-9, Pitts was determined to kill the bandit. Once again, he heard the audible tone of "missile lock," but the Foxbat's warning system must have alerted the Iraqi pilot, because he deployed another set of flares before Pitts could even fire, thus negating the missile's tracker before it even left the wing. Frustrated by the Iraqi's preemptive decoy, Pitts hastily reset the AIM-9 and fired it, only to see the hurtling missile deflected by a *third* set of flares.

By now, Pitts could tell that the Foxbat pilot was getting desperate—and Pitts himself was getting agitated. Switching back to the AIM-7, Pitts fired the missile and watched in amazement as it tracked right into the Foxbat's engines, demolishing the tail of the aircraft. "His ejection seat comes out," Pitts recalled, "and I almost hit it, going right over the top of my canopy. I never saw him separate from the seat, or a chute, but I wouldn't want to test an ejection seat at 300 feet!" Indeed, at more than 300 feet, while travelling in excess of 500 knots, it is doubtful that the parachute would deploy in time, or even sufficiently catch air to ensure a soft landing.

Tollini, meanwhile, had set his sights squarely on the second MiG. However, during the chaos of the pursuit, Tollini grew unsure if he was still following the MiG-25, or if he had vectored onto a friendly plane. "I knew there was a Navy [strike] package out there," he said. "This left me sitting barely a mile behind my target, looking at its tail, but unsure of what it was. What I *could* see was the jet's two huge burner plumes, so I asked on the radio if anyone was in burner. Having received various responses, I called everyone to get out of burner—working on the basis that if my target was indeed one of us, its pilot would comply. Well, he didn't, so I looked at him more closely, and saw that he had two missile pylons under each wing. Now I knew that it was not an F-15 or F-14. That was the moment I *knew* my target was a Foxbat. Then I started shooting."

Thrusting his F-15 to full afterburner, Tollini chased down the MiG-25, lining up an AIM-7 for his first missile shot. But his anticipation quickly turned to disgust when the missile failed to launch—"we think the rocket motor failed to light." Of the missiles carried by American fighters throughout Desert Storm, the AIM-7 was less reliable than the AIM-9 and AIM-54. Thumbing forward to the AIM-9, Tollini launched his Sidewinder missile. As he watched the AIM-9 glide away from its wing, he experienced a phenomenon shared by many other combat pilots—"time dilation"—when time literally seems to slow down during an intense situation, making the events look as though they're occurring in slow motion. Seeing the missile sail towards the MiG, Tollini felt as though it took several minutes. However, it was mere seconds before the Sidewinder met its mark on the MiG-25, puncturing the plane's underbelly and sending the Foxbat to the desert floor in a spectacular fireball.

By the third week of Desert Storm, the Iraqi Air Force was in full panic. Most of their planes had been destroyed before they ever left the ground. Many more Iraqi planes were being flown into Iran, hoping to wait out the conflict under the veil of Iranian neutrality.

As the air campaign lumbered into its second month, coalition ground forces began their assault into Iraq on February 24, 1991. Barely 100 hours after the start of the Allied invasion, however, the Iraqi Army was in full retreat and Saddam Hussein was desperate to sue for peace. Meanwhile, much of the Iraqi Air Force had been destroyed— and the few remaining planes in the air were becoming easy targets for coalition aircraft. On February 27, 1991, President Bush announced the official cease-fire. In their disastrous retreat, the Iraqis had fled Kuwait, leaving a devastated country in their wake. It would take a massive reconstruction effort to get the emirate back on its feet; but for now, the savagery of Iraq's occupation had ended. On March 3, 1991, General H. Norman Schwarzkopf met with several Iraqi generals in Safwan to discuss the terms of surrender.

Although the MiG-25 had succeeded in downing one American fighter, its performance elsewhere along the Iraqi front was disappointing. The long-awaited showdown between an American F-15 Eagle and MiG-25 Foxbat had come to pass. Like many analysts had predicted, the F-15 outclassed the Foxbat by almost every conceivable metric. Aside from its poor handling as a dogfighter, most experts agree that the MiG-25's greatest shortcoming in Desert Storm was its over-reliance on GCI stations. These ground controllers would help guide the Foxbat onto its targets, but these GCI stations had neither the reach nor multi-layered situational awareness that the American pilots had. Indeed, the avionics suite aboard the F-15 Eagle enabled it to see farther and clearer than even the best Iraqi GCI. Adding to the coalition's airborne advantage was the use of AWACS—

Sporting a rare camouflage pattern, this MiG-25R prepares to land at Welzow. At the time, this Foxbat was the only camouflaged MiG-25R serving in East Germany. (Rob Schleiffert)

flying radar platforms that could monitor the entire battlespace and relay the positions of both friendly and enemy aircraft.

Further Action in the Gulf

Following the end of Desert Storm, the United States realized that a military presence was still necessary in the Persian Gulf. A Shi'ite rebellion had erupted during the postwar chaos while the Iraqi Kurds attempted to flee the heavy-handed rule of Saddam Hussein. Thus, to protect the ethnic Kurds in the north, and the Shi'ite Muslims in the south, the US created and enforced "No-Fly Zones" over northern and southern Iraq. Citing UN Resolution 688, the United States mandated that no Iraqi aircraft could enter the No-Fly Zones—else, they would be engaged by hostile fire.

The first aerial patrols over the No-Fly Zone were dubbed "Operation Provide Comfort," (March - July 1991), followed thereafter by Operation Provide Comfort II (July 1991—

December 1996). The primary objective of both operations was to curtail Saddam Hussein's ongoing aggression against the Kurds. Provide Comfort I and II were successful inasmuch as they facilitated the withdrawal of Iraqi troops from Kurdish territory in October 1991. Thereafter, the Kurds resumed their autonomy in northern Iraq. Provide Comfort I and II were eventually superseded by Operation Northern Watch and Operation Southern Watch, ongoing missions that respectively enforced the Northern and Southern No-Fly Zones over Iraq.

It was during these No-Fly Zone missions when, on December 27, 1992, an American F-16 downed an Iraqi MiG-25. In December 1992, the 33d Fighter Squadron deployed from Shaw Air Force Base, South Carolina to Saudi Arabia for its first deployment to Operation Southern Watch. Although the 33d would be away from home that Christmas, the pilots and ground crews were motivated. Many of them had missed Desert Storm and were eager to participate in a "real-world" deployment.

During their initial intelligence brief, the pilots were informed that the Iraqi Air Force had been more active than normal along the border of the Southern No-Fly Zone. In fact, the local AWACS radar had observed Iraqi fighters "occasionally flying into the No-Fly Zone before quickly returning north of the 32d Parallel. The Iraqis initiated nearly all of the border incursions in the early morning hours, or when no US fighters were present. No provocations involving US fighters had occurred, but the activity was unusual, and pilots were warned to be on alert during their sorties."

On the morning of December 27, Lieutenant Colonel Gary North was leading a flight of F-16s on a typical Southern Watch mission—patrolling the No-Fly Zone for approximately 30-45 minutes before returning to the airbase at Dhahran. North was one of the few pilots in the squadron who had combat experience in Desert Storm; thus, he knew the airspace well and had been none too impressed by the Iraqi Air Force. "Sometimes a flight would be ordered to overfly a specific area and observe any unusual ground activity," the main purpose of these flights was to let Saddam Hussein know that coalition aircraft were still on guard.

While conducting aerial refuel, North and his comrades monitored a transmission between the AWACS and a nearby flight of F–15s. "An Iraqi MiG-25 had crossed the border into the No-Fly Zone, flown within lethal range of the F–15s, and was speeding north to safety with the F–15s in hot pursuit." One of the F–15s had visually identified the plane as a MiG-25 and requested permission to fire. But by the time he received clearance, the Foxbat had already retreated beyond the 32d Parallel.

After North and his wingman completed their refuel, they sped into the No-Fly Zone, occupying their designated flight pattern while staying alert for any lingering MiGs. Minutes later, the AWACS alerted North to a high-speed contact that had just breached the No-Fly Zone, 30 miles west of their position. The F–16s vectored to intercept, but this MiG, too, retreated northward. Shortly thereafter, AWACS reported yet another incoming bandit. This time, when the F–16s vectored to intercept, North's radar warning receiver indicated that he was being tracked by a SAM station. Yet, just as before, this MiG disappeared over the 32d Parallel, and the SAM warning indicator fell silent. By now, it was clear what the Iraqis were doing. They were sending up their MiG-25s to lure the Americans into a chase, then targeting them with SAM radars. Coalition pilots eventually started calling these tactics "SAMbushes."

As North returned to his normal flight pattern, AWACS alerted him once again. This time, a radar contact had entered the No-Fly Zone and was flying east, headed straight towards him. Determined not to let this MiG get away, North called upon his wingman to fly northward and create a "bracket"

A retired Indian MiG-25 at the IAF Museum in Delhi. Acquired in 1981, the Indian Air Force used the MiG-25 as a reconnaissance asset for its border patrols over China and Pakistan. India retired the last of its MiG-25s in 2006. (Himmat Rathore)

between themselves and the MiG-25. This maneuver essentially trapped the MiG within their immediate airspace—it couldn't retreat to the 32d Parallel without a fight. As North recalled: "Someone was going to die within the next two minutes and it wasn't going to be me or my wingman."

North requested permission to fire, having identified the bandit as a MiG-25, while directing his wingman to jam the Foxbat's radar and communications. When North heard the transmission from ground control, "BANDIT-BANDIT-BANDIT. CLEARED TO KILL," he fired his AIM-120 missile, pounding the Foxbat at a distance of 20 miles within the No-Fly Zone. "The nose and left wing broke apart instantly, and the tail section continued into the main body of the jet for one final huge fireball." It was the first confirmed kill for an American F-16 and the first kill for an AIM-120 missile.

The No–Fly Zone missions continued until the US-led invasion of Iraq in 2003. During the lead-up to Operation Iraqi Freedom, a MiG-25 shot down an American Predator drone. During the invasion itself, however, no MiG-25s went aloft. Instead, the Iraqi Air Force buried many of them in the sand—ostensibly to prevent their destruction by American firepower. One such Foxbat—a MiG-25RB—was exhumed by US forces at Al-Taqqadum Air Base in 2003. This same MiG-25 is now on display at the National Museum of the United States Air Force at Wright-Patterson Air Force Base in Ohio.

Soviet-Afghan War

While Iraqi MiG-25s battled the IRIAF, the Soviets sent their own Foxbats into neighboring Afghanistan. In 1979, the Soviets invaded Afghanistan to facilitate the socialist government's counterinsurgency against the Mujahideen. From the opening days of the conflict, MiG-25RB and MiG-25RBV variants from the 73d Air Army flew reconnaissance missions over Afghanistan. The initial detachment of MiG-25s deployed to Karshi, Uzbek SSR in January 1980. From there, they would launch several sorties into Afghan airspace.

Planning and preparation for the recon missions began almost immediately. As pilot Anatoly Doodkin recalled: "The detachment had two primary missions—ELINT [electronic intelligence] gathering along the Iranian and Pakistani border, and PHOTINT/ELINT of designated areas in Afghanistan." The PHOTINT missions typically occurred at altitudes of 21,000 to 31,000 feet. But stratospheric reconnaissance, the way it had been done over Israel, had been deemed inefficient for the Afghan theater. According to the Soviet high command, the enemy's air defense systems couldn't reach that high. And such high-altitude reconnaissance wasn't truly effective over mountainous terrain. Most missions would have to occur at lower altitudes. Daily missions were flown according to guidance from Moscow and/or the regional Air Force HQ in Tashkent. After briefing the pilots on the daily intelligence reports, the detachment commander would announce the flight schedule and the mission profiles.

First call for daily operations was typically between 6:00–8:00 AM. Within two hours, the first MiG-25 would be airborne. The number of sorties varied from week to week. Sometimes, the Foxbat pilots would fly as many as three sorties a day; other times, they would fly barely one mission per week. Recon flights could last as long as two hours depending on the priority intelligence needs. These long mission sets, however, often meant that the MiGs would land on dangerously-low fuel tanks—giving pilots "not enough fuel for a go-around in the event of a missed approach." Each pilot was also equipped with a survival kit including a Kalashnikov automatic rifle. But as Doodkin recalled, "we did not give much thought to the assault rifles in our survival kits. After all, the stratospheric altitudes we flew at, and our Middle Eastern experience [flying unmolested over Israel] made us a bit overconfident—which we shouldn't have been." For although the MiG-25s were safe from Afghan air defenses, a simple engine failure could prompt an ejection—whereupon the assault rifle would come in handy once the pilot parachuted to the ground. Moreover, the Soviet pilots received no explicit instructions about what to do after ejecting. It seemed that evasion and escape (as well as search & rescue) were little more than an afterthought. If a pilot had an in-flight emergency, however, the standing "rule of thumb" was to keep the plane airborne as long as possible and head towards Soviet airspace. If the pilot could not reach the Soviet border, he was encouraged to land at the nearest friendly airbase. "It was just as well that we had no in-flight emergencies then," said Doodkin, "but things could have turned out differently."

Doodkin also recalled his feelings of dread on every mission. Indeed, the farther his MiG flew from the base, the less assured he felt of the likelihood he'd be recovered if his Foxbat crashed. "The radio was silent as you sailed across an endless 'sea' of mountains or desert," he said. "Eventually, the course correction signals from the base would stop coming through, which meant you were 500-600 km [310-372 miles] away from 'home.' And you knew that if, God forbid, the machine should break down…no one would ever find you in that jumble of mountains."

Anatoly Doodkin flew four missions over Afghanistan. His fifth sortie, however, ended prematurely, in a curious mishap that could have cost him his life. While preparing the Foxbat for flight on the morning of January 29, 1980, a technician had neglected to properly close the gas cap on the MiG's service tank. Thus at 9:45 AM, when Doodkin thrust his engines to full afterburner during takeoff, the gas cap popped off as the plane accelerated down the runway. The leaking fuel billowed behind the plane, igniting as it made contact with the afterburners. Air traffic control watched in horror as the MiG-25 lifted off with a 130-foot trail of flames following it. The air boss ordered Doodkin to abort and land immediately. At first, Doodkin was dumbfounded by the order—he hadn't seen the flames and his instrument panel indicated everything was fine. He realized there was a problem, however, when he smelled kerosene in the cockpit. He cut the afterburners and landed moments later. The inferno had damaged the brake parachutes, which failed to deploy, and Doodkin watched helplessly as his MiG overran the runway, skidding to a halt some 1,480 feet beyond the runway. Local firefighters quickly extinguished the flames, while the negligent technician was demoted and reassigned to a less-desirable unit in the Soviet Far East.

Soon thereafter, the Soviet 40th Army determined that the MiG-25 was too big and too costly to use in reconnaissance

As the MiG-25 approaches the twilight of its service career, many of the hitherto retired Foxbats have become derelict, providing opportunistic targets for vandals and scavengers. The decrepit MiG-25RB emblazoned with the number "710" set a world altitude record in August 1977 when Alexander Fedotov flew it to 123,520 feet. (Aleksandr Markin & Alan Wilson)

missions that were better suited for lower-altitude aircraft. Thus, in March 1980, the MiG-25 detachment departed Karshi and returned to its home base at Balkhash, Kazakh SSR. As for the Afghan War, the Russians found themselves embroiled in a bloody nine-year conflict that depleted much of their military resources. By 1988, under a collapsing economy at home, and unable to crush the Afghan resistance, the Soviets conceded the fight and withdrew from Afghanistan. As a humorous aside, throughout the MiG-25's service in the

Soviet Air Force and PVO, the Foxbat was popular among the ground crews because its radar and generator cooling systems contained more than 200 liters of a potent water-methanol mix—which the Soviets could easily distill into homemade vodka. Crews had to be wary, though, as drinking too much of the coolant mix could render one blind.

Until the fall of the Soviet Union, the MiG-25 continued to serve within the nation proper, and among the satellite states in Eastern Europe. Throughout the 1980s, the R-series MiG-25s saw service in the Siberian, Trans-Caucasus, Leningrad, Central Asian, and Belorussian Military Districts. It also featured prominently in East Germany as part of the "Soviet Group of Forces in Germany"—which operated two forward-stationed MiG-25 units from the 16th Air Army. Following German reunification in 1990, these MiGs were returned to the Soviet Union.

India

The Republic of India had amicable ties with both the US and the Soviet Union. After gaining its independence from Britain in 1947, India had no stake (and little interest) in the broader politics of the Cold War. Indeed, the Indian government had more pressing matters at hand—mostly involving China and Pakistan. To this end, the Republic of India graciously accepted military equipment from both the Soviet Union and the West.

In 1981, the Indian Air Force (IAF) took delivery of six MiG-25RBK variants and two MiG-25RU trainers. At the time, the MiG-25s were the IAF's only aircraft capable of reaching Mach 3. Moreover, they were seen as critical strategic assets, capable of gathering intelligence on Pakistani and Chinese military developments—"notably armored divisions and strategic reserves close to the border." Therefore, the IAF kept its fleet of MiG-25s a closely-guarded secret. Rare was the occasion that an Indian Foxbat was seen in public.

Throughout its service, the Indian MiG-25 was nicknamed "Garuda," after the giant bird belonging to the Hindu god Vishnu. Because the IAF only had the reconnaissance Foxbats, there were no high-profile intercepts of Pakistani or Chinese aircraft. If there were any close calls with enemy aircraft (or other tales of derring-do), the Indian government has classified the details. For according to one Indian pilot, "even today we are not permitted to speak of the daredevilry these stratospheric planes have been used for. All I can say is that I more than once hit [70,000 feet] with them."

Retired IAF Air Marshal Sumit Mukerji recalled his service in the MiG-25: "Sure, you can call it an 'archaic, unsophisticated machine'. But then there was no other 'sophisticated' aircraft [in China or Pakistan] to either match its performance or shoot it down! With a navigation accuracy of (max) 10 kilometers off track over 1000 kilometers (with a lateral photo swath of 90 kilometers), strategic targets were never missed. It was amazing for its role." Having piloted the MiG-21, Mukerji and his comrades saw the MiG-25 as a welcomed upgrade. Many had described it as "an engine with a place for a pilot and some avionics."

The Indian Foxbat pilots found no cause to argue.

Mukerji continued: "It was a beast with immense power. The Tumansky R-15B engines [aboard the MiG-25RBK] each provided more than 10 tons of thrust to produce the desired performance." He was equally impressed by the Foxbat's rate of climb. Indeed, it could reach an altitude of 65,000 feet in less than seven minutes from takeoff, climbing at 100 meters per second (nearly 20,000 feet per minute). These performance metrics were beyond any plane in the current IAF inventory.

For aviators like Mukerji, the Foxbat offered a reduced cockpit workload and increased protection from Chinese and Pakistani interceptors. With the onboard navigation computer, one could engage the auto-pilot at only 165 feet after takeoff. "The Foxbat would execute the mission, photography included, and return to base (or the programmed airfield) descending to a height of 50 meters when the pilot needed to control and flare out for a landing." Throughout the mission, all the pilot had to do was manipulate the throttle settings. Although the Foxbat was detectable by radar, the Indian MiGs operated "at the highest fringes of the radar lobe, with ingress or egress (through the radar lobe) often allowing one or two blips for the radar controller to perceive. Low transition times (because of the high speed) did not provide adequate reaction time to scramble fighters; and other than a pure head-on interception with look-up/shoot-up capability…the Foxbat could survive any fighter interception."

Still, the Indian MiG-25s suffered the same mechanical and electrical problems that befell the Soviet and Arab Foxbats. But the IAF's location on the subcontinent presented its own logistical challenges. "We had problems with the tires and the fuel in the initial years," said Mukerji. "The Russians did not clear the Dunlop-manufactured tires nor the Indian Oil-manufactured aviation fuel for quite some time. We depended on the USSR for the supply of these two major items. The ongoing Iran-Iraq War curtailed supply through the Suez Canal. So, our fuel and tires would come by ship, around the Cape of Good Hope. This led to restrictions in our quantum of flying."

Throughout its service in the IAF, there were two major operations involving the MiG-25. The first occurred during the solar eclipse of October 24, 1995. The Udaipur Solar Observatory requested the IAF to photograph the eclipse,

The front view of the MiG-31 "Foxhound" (pictured here at Khotilovo Air Base) shows its commonality with the earlier MiG-25. Both planes share the same basic airframe, but with several upgrades to the MiG-31's avionics, engines, and structuring. The Foxhound was conceived as an eventual replacement for the Foxbat, and its development was accelerated after Viktor Belenko alerted Western authorities of its existence, with Belenko calling it a "Super Foxbat." Today, the MiG-31 remains in service with the Russian and Kazakh Air Forces. (Vitaly Kuzmin)

something that the IAF had never done before. "The purpose was twofold," said Mukerji—"to photograph the eclipse as it progressed, with a front-looking camera in the cockpit and secondly, to photograph the traverse of the shadow over the surface of the earth, with the belly cameras." It was a tall task, and one that required the IAF to use the two-seater MiG-25RU. The camera would thus go in the front cockpit.

After fitting the camera into place, the aircraft was raised on jacks, with its landing gear retracted to simulate flight conditions, while a "simulated sun" was built at a scaled distance to calibrate the correct alignments. "For eye protection," said Mukerji, "it was necessary [during the mission] to fly with the seat fully lowered, which posed a problem of tracking the sun accurately with respect to the camera."

Nevertheless, on the day of the flight, the two-seater Foxbat sped from the runway, right into the path of the solar eclipse. "Timings were critical," said Mukerji, "and they were met. We fed into the predicted path of the eclipse and started our photography about two minutes before the total eclipse took place. While it was getting darker by the moment, when the total eclipse took place, we were enveloped in absolute pitch-black conditions and the stars had a clarity and luminosity not seen otherwise." While those on the ground could only see the eclipse for 42 seconds, the MiG-25—flying at Mach 2.5 in the stratosphere—could see the eclipse for nearly two-and-a-half minutes. Mukerji further added that because there were no suspended particles at that altitude, "the clarity of the photographs was beyond their expectations." It was perhaps the best and clearest photograph of a solar eclipse to that date.

The next notable operation involving an Indian MiG-25 occurred in over Pakistani airspace. Since their partitioning in 1947, India and Pakistan had been fighting each other in a seemingly endless string of conflicts—mostly stemming from territorial disputes. By 1997, however, India and Pakistan had been enjoying a tenuous peace since the end of the latest Indo-Pakistan War in 1971. However, the militaries of both countries regularly surveilled the other. One day, in May 1997, an IAF MiG-25RB began what was to be a routine sortie along the Pakistani borderlands. This pilot created a furor, however, when he thrust the aircraft to beyond Mach 3 during a reconnaissance mission into Pakistan airspace. The Indian Foxbat broke the sound barrier while flying at an altitude of nearly 65,000 feet. The resulting sonic boom could be heard at various Pakistani air defense stations along the border. Indeed, had the pilot not made the sudden thrust to supersonic speed, the mission might have remained a secret. Luckily, the Foxbat's altitude and speed allowed it to escape back into Indian airspace, evading the Pakistani F-16s that had been scrambled to intercept it.

After the flyby, the Pakistani government publicly denounced the incursion, and contended that the sonic boom was a provocative jeer, trying to reinforce that the Pakistani Air Force had no aircraft that could travel at comparable speeds or comparable altitudes (70,000+ feet). India officially denied the incident, but Pakistan's Foreign Minister claimed that the Foxbat photographed strategic installations near the capital of Islamabad.

Due to the growing scarcity of spare parts, and India's acquisition of UAV drones, the IAF retired the last of its MiG-25s in 2006.

★

Epilogue
Foxbat Sunset

A t this writing, the MiG-25 Foxbat remains in service
with only a handful of countries. Even within these
countries, few Foxbats remain airworthy, and many are being
grounded involuntarily due to lack of spare parts. Given the
current operational rates, no Foxbats are likely to remain
in service beyond the year 2025. Even the Russian Air
Force (the MiG-25's legacy operator) has forsaken its fleet
of Foxbats in favor of the MiG-29, MiG-31, and the ever-
growing stable of Sukhoi fighters and interceptors. Given the
cost of maintenance and upgrades, the MiG-25 is no longer

A MiG-25 monument at Dubna Air Field, Moscow. (Oleg Novikov)

cost-effective for any fleet in which she flies.

From its unveiling in 1967, to its infamous defection in Japan, to its service in the "hot spots" of the Middle East and Central Asia, the MiG-25 has proven itself to be a resilient and adaptable interceptor. Although ill-suited for dogfights, and not well-engineered by Western standards, the Foxbat fulfilled a critical role for the air forces in which it served. For the latter-day Soviet Union, it was an interceptor and reconnaissance platform that could fly alongside even the fastest Western aircraft. Its presence alone prompted a renewed vigor in the development of the F-14 Tomcat and F-15 Eagle, the very planes that would decisively outclass her in combat. The Egyptians (although never permitted to have a Foxbat of their own), marveled at the speed and accuracy of the MiG-25's PHOTINT capabilities. For the Israelis, the Soviet Foxbat was an elusive bandit that not even their best interceptors could catch. For the Syrian Air Force, however, the MiG-25 provided little more than fodder for the Israeli F-15s. For the Iraqis, it was a solid reconnaissance platform and, although frequently downed by Iranian F-4s and F-14s, the interceptor variant did score a few victories against the IRIAF. And for the Republic of India, the MiG-25 was a welcomed asset for its border security against the likes of China and Pakistan.

Although the Foxbat had its share of mechanical problems, and numerous losses in air combat, the MiG-25 had one qualitative strength that was beyond reproach—its speed. During a time when airspeeds of Mach 3 were deemed unreachable for fighters and interceptors, the Foxbat broke records and shocked even the most seasoned of Western analysts. The speed with which the Foxbat could break away from pursuers (whether over Israel or Iran) was its greatest defense. Although the Foxbat is undeniably in its final years, there can be little question regarding its prolific legacy as one of the fastest interceptors and reconnaissance jets the world has ever seen.

Select Bibliography

Aloni, Shlomo. *Israeli F-15 Eagle Units in Combat*. London: Bloomsbury Publishing, 2013.

Barron, John. *MiG Pilot: The Final Escape of Lieutenant Belenko*. New York: McGraw-Hill, 1980.

Brown, Craig. *Debrief: A Complete History of U.S. Aerial Engagements 1981 to the Present*. Atglen: Schiffer Pub, 2007.

Cooper, Tom, and Farzad Bishop. *Iran-Iraq War in the Air, 1980-1988*. Atglen: Schiffer Publishing, 2002.

Cooper, Tom, and Farzad Bishop. *Iranian F-14 Tomcat Units in Combat*. Oxford: Osprey Publishing, 2004.

Cooper, Tom, and Albert Grandolini. *Libyan Air Wars: Part 1: 1973-1985*. Helion and Company, 2015.

Cooper, Tom, Albert Grandolini, and Arnaud Delande. *Libyan Air Wars: Part 2: 1985-1986*. Helion and Company, 2016.

Dildy, Doug, and Tom Cooper. *F-15C Eagle vs MiG-23/25: Iraq 1991*. London: Osprey Publishing, 2016.

Gordon, Yefim. MiG–25 *Foxbat and MiG-31 Foxhound: Russia's Defensive Front Line*. London: Aerofax Midland Pub, 1997.

Gordon, Yefim. *Mikoyan MiG-25 Foxbat: Guardian of the Soviet Borders*. London: Midland Pub, 2007.

Gordon, Yefim. *Soviet Spy Planes of the Cold War*. Barnsley: Pen and Sword, 2013.

"The Last MiG–25 Foxbats of the Syrian Arab Air Force." *The Aviationist*. March 2, 2018.

Mladenov, Alexander. *Soviet Cold War Fighters*. Stroud: Fonthill Media, 2017.

Morse, Stan. *Gulf Air War Debrief*. London: Airtime Publishers, 1991.

"Say Goodbye to Russia's Mach-3 Spy Plane." *The National Interest*. April 19, 2019.

Sweetman, Bill, and Bill Gunston. *Soviet Air Power*. London: Crescent Books, 1978.